GEOGRAPHY

IDEAS IN PROFILE
SMALL INTRODUCTIONS TO BIG TOPICS

GEOGRAPHY

DANNY DORLING

&

CARL LEE

Maps by Benjamin D. Hennig

P

PROFILE BOOKS

First published in Great Britain in 2016 by
PROFILE BOOKS LTD
3 Holford Yard
Bevin Way
London WC1X 9HD
www.profilebooks.com

A CIP catalogue record for this book is available from the British Library.

ISBN 978 1 78125 530 8

eISBN 978 1 78283 196 9

Maps by Benjamin D. Hennig, University of Oxford
Text design by Jade Design *www.jadedesign.co.uk*

Printed and bound in Italy by L.E.G.O. S.p.A.

CONTENTS

We would like to dedicate this book to the thousands of geography students who, over more than two decades in classrooms, lecture theatres, field trips and tutorials, have helped us learn so much more about a subject that we love than we could have ever learned alone.

ACKNOWLEDGEMENTS

As always, we stand on the shoulders of others (or at least look over their shoulders) for all our ideas. Maybe one day we will have a big idea of our own, but we doubt it. Particular thanks must go to Michael Bhaskar, who was able to understand the original idea for this book, which was conceived on the Île de Ré in the summer of 2014 and finished in the Jura Mountains in the summer of 2015. Thanks also to Mike Jones and Cecily Gayford for editing the text and, before it reached that stage, Grant Bigg, Tony Champion, Paul Chatterton, David Dorling, Alan Grainger, Sally James, Karen Robinson, Natasha Stotesbury, Laura Vanderbloemen, Jenny Watson and Robert Whittaker, who all provided invaluable commentary, much of which we heeded and some we possibly should have but did not.

An inspiration to us both is the work of Benjamin D. Hennig, and we thank him for permission to use his innovative cartographical skills in producing all the new maps included herein.

MAPS

INTRODUCTION

'What you know you can't explain, but you feel it. You've felt it your entire life, that there's something wrong with the world. You don't know what it is, but it's there, like a splinter in your mind, driving you mad.' – Morpheus.

The quote is from *The Matrix*, a movie, not reality ... or is it?

On 25 January 2015 the MSC *Oscar*, a Panamanian flagged ship laden with goods, set sail from the port of Dalian in China. Sailing southwest, it picked up yet more cargo at one of the world's largest ports on the southern tip of Malaysia, Tanjung Pelepas. Next, cruising at a sedate 26 kilometres an hour, it passed through the Malacca Strait, the Suez Canal and the Strait of Gibraltar. The entire voyage to Europe took just over five weeks. On 3 March 2015, when it arrived in Europe at Rotterdam's huge container port, the ship made the news. It was nearly 400 metres long, 60 metres wide and 73 metres high. It could transport 13.8 million solar panels or 1.15 million washing machines or 39,000 cars. The MSC *Oscar* was the world's largest container ship.

It almost certainly isn't the largest container ship today. When the MSC *Oscar* was launched keels were already being laid down in the shipyards of South Korea for ships that will by now have surpassed even her 19,224-container load. A container is often referred to as a 20-foot equivalent unit (TEU), a 'unit' designed to be hauled by a lorry. Try to imagine almost 20,000 lorries travelling end to end. Even

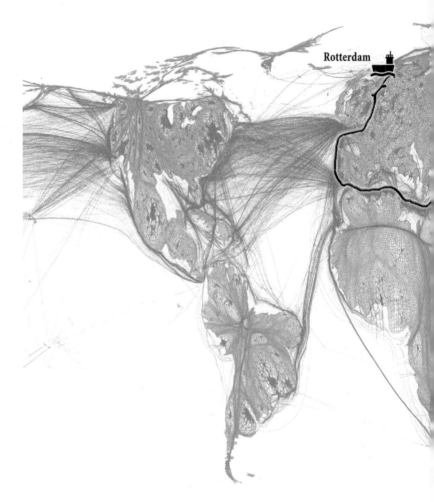

Rotterdam

This map of the world has been resized to represent where humans currently live on earth by giving each person equal prominence. The largest cities can be seen, and deserts and the polar north almost disappear. This map also shows how the world is connected via the shipping lanes, flight lanes and underwater cables that

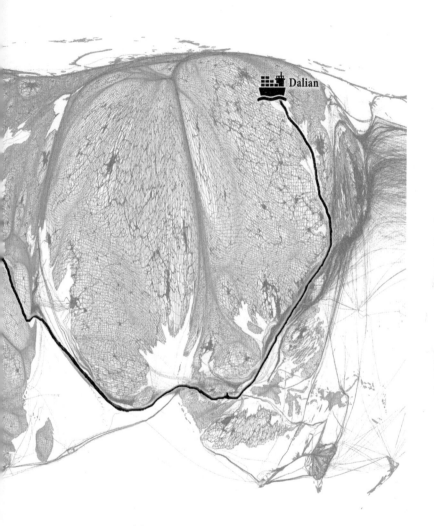

Dalian

carry most of the trade that drives the global economy. The route
of MSC Oscar's first voyage from Dalian, China, to Rotterdam,
Netherlands, via Tanjung Pelepas, Malaysia and the Suez Canal
is shown. This journey took 36 days, with the ship passing close by
nearly half the planet's human population in that short time.

closely packed in a traffic jam, the cavalcade would be over 160 kilometres (about 100 miles) long. The MSC *Oscar* is the physical embodiment of a globalised world. Its much-heralded fuel efficiency and relatively slow cruising speed are signs of recent greater concern over the environmental impacts and costs of transporting the multitudinous manufacturing bounty of China and its neighbours for the 12,569 nautical miles that separate Dalian from Rotterdam.

The value of goods transported from China to Europe is an integral part of an equalising movement of wealth from West to East that has been one of the most dramatic economic developments of the early years of the 21st century. Goods flow from East to West, but now far more money than ever before has to flow back in recompense. The wealth of China continues to rise more rapidly than that of most of the rest of the world, while that of almost all of Europe has fallen in recent years. Yet still the ships come – ever newer, bigger and better – full to the brim with the stuff of consumption.

The MSC *Oscar* tells us something about globalisation, sustainability and equality. These are the three key themes of this book because they are the key themes of geography in the 21st century, and our concern here is with exploring what the study of geography means today and what it could mean for the future. The academic subject of geography has existed for many years, but the themes of globalisation, sustainability and equality have not always been of such paramount interest to those who pursued the study of geography in the past.

Three hundred years ago, at the beginning of the 18th

century, trade between Europe and China existed but was limited, largely taking place in the highly taxed and regulated port of Canton on the Pearl River delta. Britain was about to become the first place on earth to move towards an economy powered by fossil fuels. Before they were powered by steam generated by coal (then oil), any ships plying the Europe to Canton route relied on the power of the wind and would have taken between four and five months to make the journey. They also had to round the Cape of Good Hope at the southern tip of Africa rather than pass through the not-yet-dug Suez Canal.

Since antiquity geography has been the subject of measuring and marking space in the world. It rose to the academic fore through the exploration of new lands and as a route map and resource inventory for colonial endeavour. Such priorities helped set in motion an increasing level of global inequality, driven forward by a cornucopian worldview. From such a perspective it was one's duty to take possession of 'God's bounty', which appeared, at one time, to be limitless.

Before European colonial aspirations were unleashed some 500 to 600 years ago, people had little idea of where the limits of the earth lay. Cartographers – those who drew the maps, the vast majority of whom were Chinese, Indian, Arabian or European – wielded great power and held carefully protected secrets about trade routes and coastlines. Geography was about territory and mapping territory so that you could control land.

Shortly before Columbus set sail in 1492 on his first attempt to set foot in the Americas, a revolt against the dynastic Habsburg empire of Maximilian I by Flemish cities, including Rotterdam, was crushed. It was into the port of Rotterdam that MSC *Oscar* would sail some 523 years later. History matters to geography. Both subjects complicate each other, and so we begin this book with background material from quite some time ago. Geography is not history; but history, along with much else, helps us to understand geography.

So what is geography? What is geography today? One answer is that without geography we can't explain the world around us. Geography is all around us. It is about what is where and where is what; and why and when, and who and how. It is about exploring places and spaces. Almost everything has a geographical dimension. Nothing lies outside the purview of geography because everything is connected to everything else. To many people, that is not a particularly helpful answer, but, because geography is not a traditional academic discipline in the way that philosophy, economics or politics are, it has always proved an elusive subject to pin down to a simple, neat definition. In attempting to make it a neat discipline, geography is sometimes reduced simply to being concerned spatially about phenomena – that is, in the way that history has a temporal focus, geography has a spatial focus. Although 'space', and by association 'place', are clearly central ideas in the study of geography, the subject is also about so much more.

Geography is about the planet we inhabit, from the water that gives it life to its extensive biodiversity. But crucially it is about the energy that courses through its myriad environmental systems. Without an understanding of energy you cannot understand geography. This is the energy that builds mountains and then destroys them. It is the energy that flows through our atmosphere daily, bringing us our weather and, in the longer term, the energy fluctuations that alter our climate. It is the energy that gives us the food to sustain ourselves. It is the energy, stored in fossil fuels and formed from the decomposition of prehistoric plants and animals, that has enabled us to put the spark of electricity into our lives. Geography helps explain how all of this energy has transformed who we are and how we affect our environment and how our environment affects us.

Geographers have a tradition of being curious explorers of both places and ideas. Where does that highway go? Who lives in this big house and why? How did we arrive at where we are? When are we going to learn to live together? Can you really consume more and more and does it really make you feel better? Is there something nagging away at you every day – a splinter in your mind – a thought that somehow there must be a better way to live? Questions abound.

Geographical questions are never stand-alone ones. All the questions we ask lead to other questions. Geography is about joining up the dots that help make up the big picture. Connections are everywhere. The distinction between human and physical geography is often a false dichotomy: the two are intimately connected, the unifying factor being the energy that flows through all that we do, see and know.

Geographers need to know a little biology and chemistry, sociology and politics, and some mathematics and economics. A few languages can come in handy, too, as languages tend to be quite geographical. But geography specialises in doing something no other academic subject does so well – it looks forward.

What has propelled geographical knowledge into the forefront of thinking about the future is the acknowledgement that we, humanity, have pushed our planet to the edge of an environmental catastrophe and that we have achieved this in a relatively short space of time. To some this may seem a bold assertion; to us, it is increasingly evident from the answers we get from the questions we ask. When it comes to the environment, everywhere is connected to everywhere else. Your car exhaust really is connected to the health of tropical forests in more ways than you can probably imagine. Many electrical devices you own have within them tantalum capacitors, which are produced from coltan, a rare earth mineral, that is predominantly mined, unsustainably, from central African nations, often fuelling conflict and the destruction of tropical forests. The fate of coral reefs in the vast Pacific Ocean sits on the shoulders of the billions who inhabit the world's increasingly densely populated cities. Connections abound.

We can look to a probable future of 10 billion of us, with rising inequality between the haves and have-nots. We predict a declining level of global biodiversity and squeezed biological and geological resources to sustain us. What's more, we see an economic system whose advocates still do not understand simple concepts. Average incomes will not grow forever simply from some trickle-down of wealth from those who benefit via the compound interest on savings. Disasters unfold when greedy individuals maximise their short-term gains unfettered.

We expect that our children will live in a world in which natural disasters will be increasingly devastating, in which rising sea levels will remove small island states from the map and in which the glaciers, whose snowmelt sustains millions, will be progressively diminished. But we also realise how much we did not know in the past. Only recently have we realised that glaciers were once much larger. It was not even fully accepted that continents continually moved, albeit slowly, until the latter half of the 20th century. Until recently it was thought that another ice age was imminent. It is remarkably easy to get the future wrong.

There must be so many things we do not know today that in a few years' time will appear obvious, but we can be absolutely certain that never have we known so much about our world. Never have so many humans lived so long. Never have so many people consumed as much as we consume collectively today. Never have we been better connected, with the geography of distance shrinking us to the geography of the Internet's 'right here, right now'. Never have we been more in touch with the multitudes who make up humanity and the environments that sustain us. In so many ways this is the best of times, but it is also the worst of times.

How will we find our way through the myriad and complex routes ahead of us, the numerous alternative tracks that crisscross the map of our future? This is the central role of geography in the 21st century. Geography is the subject closest to what is just about to be, what is just about to happen. Geography really should be, and often is, the splinter in your mind, driving you mad. Once you begin to see how everything is connected to everything else there is no turning back. You feel it. You've felt it your entire life. And once you start to look you can very clearly see that there's something very wrong with the world, in fact many things. Then, when you learn more about geography, you begin to know what is wrong but not necessarily what to do about it.

This book is a collection of short stories that strive to explain the role of geography in understanding how we got to where we are now and where we are going. These stories are, of course, our take on geography, and many others would not agree with this selection. Some people have said to us 'you should have included …', 'what about…?', 'I can't believe you did not consider…'. As we have acknowledged, geographers make mistakes, and we will have made some too. If you are reading this book not in 2016 but a few decades in the future, you may be amazed at what we missed out and you will be puzzled or amused by what we included. You may also struggle to understand why so much that seemed so important to us today was not quite what we should have been concentrating on. We know we won't be looking at exactly the right things, but we also know that we now look at the world very differently from the way in which most geographers just a generation before us did. The academic subject of geography is changing as the planet changes.

The stories in this book begin by looking at the historical tradition in which geography came to mean something as an academic subject and then started to come of age as one of the key disciplines to study. This period, the millennia to 1800, we broadly term 'the organic age'. We then turn our attention to globalisation and how it evolved, after 1800, out of the industrial revolution at the beginning of the 'fossil fuel age'. Globalisation has become a process that unites us all. Because, like all geographers, we take the long view, we then consider how and why we should think about sustainability. In our penultimate chapter we consider the role that a fairer, less unequal world would play in achieving a more sustainable world. The final story in this short book of many stories gazes into the future – the 'post fossil fuel age'. In all these chapters we unfurl our maps to show how geography remains an essential tool for not just describing and visualising the world in which we live but the explanatory tool that is essential to making our world a better place for all.

1

TRADITION

In its many different international incarnations the TV programme *Who Do You Think You Are?* has captured the imagination of millions of viewers worldwide. The human histories each episode weaves rarely go beyond a couple of centuries, but each of the ancestors' stories seems to contain some surprises. Modern science has to some degree laid bare the biological history – and mystery – of all of us, and we now know from where we all come, but just a dozen generations ago evolution was not even a thought, we really had no idea who we were, although many claimed to have such knowledge. Creation myths were common.

It is easy to feel small in a vast universe, especially when you look up at night to a sky full of stars and ask, 'who are we?' Yet humans are now the dominant species on the planet. All we now really have to fear are tiny bugs, viruses, bacteria and each other.

Geographers use many clues to try to better understand the world. Some, like stars, are extremely large clues; one points north and has guided us for centuries, but they all also hint at the origins of the universe. Other clues are so small that they can only be seen through an electron microscope, such as mitochondrial DNA.

It is partly the insights that we have gained from the study of mitochondrial DNA that have enabled us, so

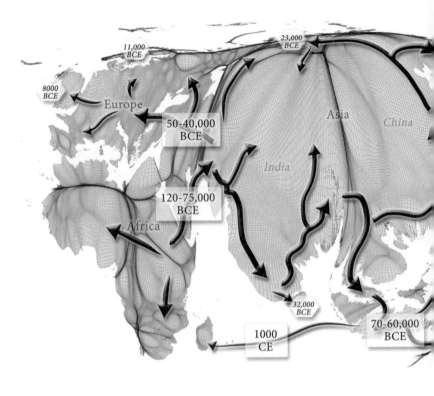

This map illustrates the migration of humanity across the Earth, with all movement originating in Africa and with the estimated dates of arrival shown at various locations. It is drawn on a base map that reflects the population density of humanity today. However, that density, with most humans now living in Africa,

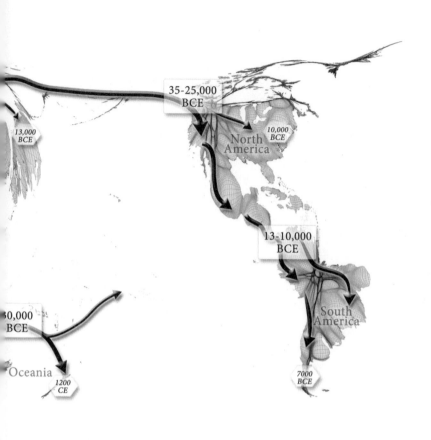

India and China, has been long established, and so this base map is similar to one showing area in proportion to all humans that have ever lived. In such a map Australia and the Americas would be drawn even smaller than those continents are drawn here, and Europe would be a little smaller.

recently, to plot and map the spatial evolution of humanity across the Earth. Following a matrilineage (mitochondrial DNA is inherited only from mothers), the journey of humanity from its 'out of Africa' moment somewhere around 100,000 years ago can be tracked with far greater certainty than was, again, only recently possible. We used mainly to make guesses from the spread of artefacts. Now we still guess, and estimates can vary by tens of thousands of years, but we are beginning to draw the global map of who we all really are, where we come from and how long it took us to get to where we are now.

At the original source of our DNA trail is a 'mitochondrial Eve', a woman who would have had at least a daughter who went on to have a daughter of her own and from whom all of humanity is related exclusively through the maternal side. Scientists continue to haggle over when such an individual existed, but not if she existed – the range of estimates for her lifetime is between 90,000 and 160,000 years ago. They are far more certain of Eve's location: the plains of east Africa, the point of human origin on the map on the previous two pages.

Genetic insights allow us to map the progress of humans across the planet's surface, facilitated by environmental change as ice sheets retreated, but also to relate these mass- and sometimes micro-movements to the major changes in environmental conditions that human expansion brought to different places. Humans themselves changed slightly as they shaped and were shaped by their surrounding lands. They also began to change the plants and the animals around them through selective breeding. Our genes and our geography are intertwined.

In some areas human skin lightened; in other places some of us became able to drink other mammals' milk due to genetic changes. But these were all minor adaptations. Ever since mitochondrial Eve, the biological commonalities that exist throughout humanity have meant we are one race – the human race, *Homo sapiens*. We are all related. The planet is our common home.

Considered as the history of the single human race, the evolution of humanity across the planet allows us to contextualise historical epochs. From the point that humanity began to evolve culturally from hunter-gathering to a settled, agricultural existence some 12,000 years ago through to the late 18th century when the industrial revolution picked up momentum, this was an epoch that can now broadly be seen as having been the organic age. In this period human endeavour is framed by the ability to put natural or organic sources of energy to productive work.

In the organic age our energy sources were primarily biomass, biological matter from living or recently living plants and animals, and the harnessing of on-going biophysical processes, such as water flow, gravity, wind, ocean currents and fire. This chapter, Tradition, is concerned with this organic age of geographical development. Although this period may be simply thought of as history, it is also the age by which we can judge the relative impact of humanity upon the Earth today. The organic age was a period, after the last ice age and particularly since 8000 BCE (Before Common Era), when global temperatures and climate were relatively stable. It was within this 'Goldilocks' moment in geological time – not too hot, not too cold – that humanity asserted itself across the world, drawing on the natural, renewable energy resources that could be discovered and explained to others, as language and knowledge expanded.

A common focus when examining this epoch is to enquire why so many of the world's earliest 'civilisations' sprang out of the land where three continents – Europe, Asia and Africa – meet. Indeed, why did this area become the cradle of successive monotheistic religions that are intent on propagating a specific perspective to understanding life on Earth and that today claim over 50 per cent of the global population as adherents?

There is only one point on Earth where three continents meet. However, to claim that that is the reason so many civilisations began there (without better reason) can be seen as what some call crude 'geographical determinism' – the idea that the human condition (and character) is mainly determined by the environmental conditions in which it evolves. The bio-geographer Jared Diamond has explored whether such a deterministic perspective is valid in his book *Guns, Germs and Steel* (1997), and he deduces that it is only partially relevant, with advantages being drawn from the domestication of animals and plants and the development of pathogen resistance that settlement where the three continents met led to over time. Diamond is careful to not attribute such advantage to something genetic within strands of humanity – the guns part of this thinking arose through European avarice and expansionism, the germs were Old-World germs, and the steel was first made in Africa and then China.

Rather than trying to guess why one particular continent

came to hold so much power at any one point, it can be more interesting and informative to consider the overall story of how humans gained so much power over other animals and plants, through utilising energy, and how in turn that use of energy has now altered the planet.

As the ice sheets began their retreat after the last glacial maximum about 20,000 years ago, a huge remodelling of the shorelines of the global land masses occurred. This process, known as eustatic change, is influenced not simply by the release of locked-up water from terrestrial glaciers raising sea levels but also by the higher water temperatures that expand the space occupied by the oceans. Slightly warmer waters take up much more space. Land also rises (isostatic change) when not weighed down by ice, but most of the world's land surface area had been ice-free even at the glacial maximum.

With a rise in sea levels of around 120 metres since the last glacial maximum, land that supported some of the earliest areas of human settlement, such as the Persian Gulf, was inundated. This pushed the most fertile zones further northwards up the Tigris and Euphrates River valleys. Today these rivers bisect modern-day Iraq.

Melting ice also swept away the Bering land bridge – between what today we call Alaska and Russia – closing the route by which humans moved from Asia into North America and thence swarmed down to the southern toe of the Americas at Tierra del Fuego. Similarly, people's migrations down the Malay archipelago were easier when sea levels were lower and before many potential stopping-off-points had sunk under the rising waters. Rising oceanic waters cut off previously translocated populations, such as the aboriginal populations of Australia, and the surging North Sea inundated the low-lying lands that connected Great Britain to the mainland of Europe.

When Paul Crutzen, the Nobel Prize-winning atmospheric chemist, popularised the term 'Anthropocene' in the 1990s he was attempting to illustrate the enormous human impact on the natural environment that humanity had been responsible for since the beginning of the industrial revolution. The word Anthropocene simply means the geological era dominated by the intervention of people. Crutzen and his colleagues saw the transition from an organic (biofuel) energy economy to a fossil fuel economy within some European nations during the late 18th century as the time when, unequivocally, the human impact on the environment became global in its scope.

> To assign a more specific date to the onset of the 'Anthropocene' seems somewhat arbitrary, but we propose the latter part of the 18th century, although we are aware that alternative proposals can be made (some may even want to include the entire Holocene). However, we choose this date because, during the past two centuries, the global effects of human activities have become clearly noticeable.
> (P. Crutzen and E. F. Stoermer, *The 'Anthropocene'*, IGBP newsletter 41, 2000)

Some observers have suggested that human activity set in train fundamental changes in global environmental systems as far back as 8,000 years ago. This 'early Anthropocene theory', most famously propounded by the American geographer William Ruddiman, is challenged widely.

Ruddiman's early Anthropocene theory suggests that the human impact on the environment started to become significant from the moment that hunter-gatherers made the transition to agriculture with the clearances of original habitat and the adaption of species, both plant and animal, for human advantage.

In order to support his theory, Ruddiman pointed to measurements of methane concentrations in the atmosphere that rose – rather than fell as long-term atmospheric models suggested they should have – from the point when people began to settle on the land in large numbers. He argued that the rapid adoption of rice farming in East Asia (and the weeds that grew with the rice) and the use of fire to clear swathes of forest across Europe and the Americas contributed to this early rise in methane levels in the atmosphere above what might be expected without anthropogenic – human – intervention.

If Ruddiman is correct, then what he is describing is the first example of humanity capturing energy and putting it to work on a scale significant enough to create a degree of human-induced disorder in the biosphere. In the early days it was the burning of forests and the conversion of the sun's energy into food, such as rice, maize and corn, that upped the amount of energy humans could and did utilise.

One of the most momentous changes brought about by the rise of settlement through the successful production of local agricultural surpluses about 7,000 years ago was the ability of humanity to think beyond simple subsistence. This was the advantage that was brought about by the more efficient utilisation of nutritional and material energy resources. The world became for some a place to ponder, to reflect upon, to paint and, eventually, to write about. Painting and ponderings had been around before settlement, as many Neolithic cave paintings testify, but increasingly, from the beginning of the early Anthropocene, pondering began to be about a particular place that was increasingly known as 'home'. A sense of place was probably the first geographical thought.

Having time to think more brought about innovation and new technologies. Things we never knew we wanted became essential – the plough, for example – and things we could not produce ourselves came to be exchanged or traded. Humans started to become increasingly obligated to each other beyond their immediate kith and kin: to be in each other's debt.

As humans congregated into larger groups, rules, regulations and religion developed for new orders to exist and more complex organisation to be realised. Controversial evolutionary biologist Richard Dawkins has suggested that religion can confer an evolutionary advantage to some societies. Power and control became an increasing imperative for the formation of complex, spatially concentrated populations – or cities, as the more plain speaking might call them.

Damascus, today the capital of Syria, is one possible candidate for the longest continuously occupied city in the world. Certainly, settlement in Damascus stretches back 8,000 years, and over that period of time it has been a hub for humanity. It is mentioned in Genesis, the first book of the Bible, which was probably written around 3,500 years ago. Abraham's steward Elizier hails from its already well-developed environs, and later, probably only 1,900 years ago, in Acts the souks and streets of Damascus are eulogised: 'And the Lord said to him, "Get up and go to the street called Straight: and inquire in the house of Judas for one called Saul" ' (Acts of the Apostles 9:11).

In Damascus, between 2,000 and 3,000 years ago, a form of early capitalism based around trade and controlled by merchants was beginning to thrive. Coming into the city from the east were silks and spices; from the west came metalwork and grain; from the south came slaves and salt; and from the north came horses and more slaves. And if you were trading, or for that matter calling on Saul, the place to go to was Straight Street – the Roman main street of Damascus – which at that time was one of the greatest marketplaces in the world.

Outside the city walls of Damascus in 1 CE (Common Era) would have been thousands of camels, which would have been the means of transport for both people and goods across vast distances, taking weeks, often months, to reach their destinations. A caravan of camels from Damascus would take just under half a year to reach China, travelling approximately 40 kilometres a day, bandits, water and natural hazards permitting. It is easy to see why some traded goods were so expensive: at that time transport costs were extremely high, and losses of goods, people and camels were common.

The cities of early history were places for the exchange of ideas, philosophies, people and things. Many of these cities waxed and waned over time. No city better demonstrated this than Rome where, by 200 BCE, its million plus residents – if we include the 350,000 slaves – required thousands of tons of wheat, 30 million litres of olive oil and 75 million litres of wine a year to survive in the manner to which they had become acquainted. Lead came from Britain, silver from the Black Sea coast, salt from the Sahara, wheat from Carthage, slaves were traded up the Nile from the heart of Africa, spices were brought from India and silk from China, carried across the vast plains and mountains of Asia. In 100 CE the appropriation of wealth from Rome's dominions and beyond ran to a total 'take' that today would equate to $97 billion a year (at 2015 prices). Rome sat at the centre of the Old World's first partially global trading system.

The consumption generated by as large a population as ancient Rome's resulted in huge amounts of waste. The best-known Roman rubbish dump was Monte Testaccio, a site on the edge of Rome, where 50 million smashed olive oil amphora (containers) remain in a mound 35 metres high. The need of Rome's vast population to draw on the energy reserves of elsewhere, be it wheat from Carthage or slaves from all over the Roman Empire and beyond, became increasingly problematic and was possibly part of the explanation for the empire's eventual decline.

Moving grain, ivory and lead around from where they were sourced to where they were consumed required a huge amount of energy. All of this energy was either captured from natural sources, such as the use of wind power in ships' sails, or generated by feeding human beings and requiring them to row or trek with pack animals (which also had to be fed to power the animals' legs and lungs).

Fleets of ships plied their trade across and around the Mediterranean Sea, roads radiated across occupied lands, camel trains plodded across desert sands – all forms of transport utilising different sources of energy to move things from there to here. And what things! For the lucky few at the heart of the Roman Empire these were riches beyond the imagination of anyone who had ever lived in any society up to then.

The most famous of Roman geographers was Ptolemy, who lived from about 90 to about 168 CE, during the height of imperial Rome. In his most famous work, *Geographica*, he constructed a map of what was (to the Roman Empire) the known world. It stretched across the mid-latitudes from the Azores islands in the north of the Atlantic to central China, from Scandinavia to the impenetrable Sahara Desert and then across to the mysterious Nile, a river with no known origin.

The Romans did not explore sub-Saharan Africa, nor did they have any idea of the Americas or that Australia existed, just as people in Australia knew nothing of Europe or of how their lands would later be named. Ptolemy knew he had covered only a quarter of the Earth's surface – theoretically, he accepted that the world was a sphere – yet even the technology and energy resources of the mighty Roman Empire could not project across the Atlantic Ocean or even far into Asia, at least not to anywhere beyond China.

In Rome's success were sown the seeds of its decline. 'Barbarian' tribes from the north finally deposed the last Roman emperor in 476 CE, but a multitude of domestic factors had weakened Rome to the point it could no longer resist its hostile northern neighbours: the fading of the imperial system of trade and tribute, an inability to maintain transport connectivity, a declining ability to extract yet more energy from the land and possibly even an over-reliance on poisonous lead.

Rome's decline was spectacular. By the time Michelangelo painted the Vatican's Sistine Chapel in the first decade of the 16th century, Rome's population had fallen from over one million to 60,000. It had been hovering around the 50,000 mark for the previous 1,000 years. The monuments of the Roman Empire are today widely scattered around the shores of the Mediterranean and can be found stretching northwards as far as to Hadrian's ultimately inconsequential wall.

At the same time that Rome and Europe were declining, to the east China was rising. The actual size of the global economy did not fall with the decline of the Romans. Instead, in China resources and power grew, and so worldwide a balance remained with the disparate powers of what is now India lying between and being buffeted by both sides.

In 100 CE China was a land that was increasingly being unified under a central state. Chinese historians often cite the Han dynasty of this era as a 'golden age', but this was only one historical era, or dynasty, in the continuing development of the Chinese state up until the end of the organic age. The dynasties came and went after the Han's demise in 220 CE – Jin, Sui, Yuan, Ming and Qing are an incomplete selection – but the idea of a central Chinese state and distinct geographical culture remained throughout this progression of dynasties. In the later 19th century Franklin Hiram King, the father of 'soil physics' in the USA, visited the great Northern Plain of China, where he saw peasants tilling and fertilising the land in a way that he estimated, perhaps a little over-enthusiastically, had not altered for 4,000 years. Certainly, there was much that the more recent (by a couple of millennia) Han dynasty would have recognised in the way land was put to productive use.

What King had discovered was that it is possible to coexist with nature in a sustainable way, even with high population densities and even in a peasant society. A highly productive agricultural infrastructure enabled China to support cities with a higher population density than anywhere else in the world. In China, from 600 CE, a developed network of urban centres that enhanced each other through trade began to evolve. It would be almost a millennium before Europe could match this.

In 618 CE the Chinese Tang dynasty set in motion an age of great transformation and technological advancement. The capital of this new empire was Xian, in present-day northwest China. The Tang dynasty traded widely with the rest of Asia and also expected 'tribute', the idea that wealth was transferred to the centre as a mark of respect – you might think that taxation is similar, although it is not necessarily given with any respect today. This tribute was similar to that extracted by the citizens of Rome from its wider empire. With increasing resources and prosperity, the population of China grew from 49 million in 630 CE to 80 million by 900 CE, and an increasing proportion lived in urban areas.

Xian was the gateway to the great trading 'Silk Road' into Central Asia. It became by 700 CE the largest city in the world, with a population hovering around the one million mark. It has been suggested that in the age of organic energy one million was close to the upper limit that could be supported in an urban area given the transport technology available at the time, the size of the agricultural hinterland that could be harnessed and the availability of fuel energy from forests in close enough proximity. It is certainly true that not until London's population rose beyond 1,350,000 in 1825 was there a global city bigger than the largest city in China (in 1820 that had been Beijing, with a population of 1,300,000). Modern London was the first great city of the fossil fuel age.

China also developed a significant cartographic idea of geography – place, people and production – that became increasingly useful for organising a complex centralised state. The first maps using a gridded layout and scale were developed in China in the 3rd century CE by the cartographer Pei Xiu. The production of maps reached new levels of sophistication in China, although only fragmentary remnants of many early artefacts of geography remain from which we can learn this.

Rumours of China's existence occasionally filtered through to Europe, along with rare and highly prized spices, pottery and maps. We know much today through the documented accounts and activities of groups of merchants, such as the Radhanites, Jewish traders whose business, with the help of camels and ships, occasionally connected China and Europe. After the Tang dynasty fell in 908 CE the

Radhanite and similar networks also collapsed and world trade subsided.

China did not look far beyond its prescribed territory apart from a brief flirtation with wider exploration in the 15th century, when the Chinese admiral Zheng He set sail on the first of his seven voyages of discovery, diplomacy and commerce. The 15th-century Ming period of Chinese history was relatively stable, and few external threats challenged the empire, so the Chinese, temporarily, looked outwards. They made it all the way to the east coast of Africa – to which both Chinese coins and DNA testify. Their merchants travelled as far west as Mecca, and they developed a mutually beneficial commercial relationship with the southern Indian port city of Calicut, a place the Chinese much admired for its accurate use of measurement and effective business regulations.

The Chinese ceased explorations only when the pressures of fending off northern 'barbarians' used up too much of their spare energy, their 'surplus' resources. Instead, they turned their attention to building up the Great Wall of China – a massive defensive structure behind which an insular Chinese polity developed a more inward-looking society that shunned further exploration. A disproportionately large amount of China's energy resources were expended on not only constructing the Great Wall but also maintaining it.

The brief trading relationship between India and China that Admiral Zheng He forged predates the arrival in India of Vasco da Gama, the Portuguese explorer, by over 90 years. Yet it was those later maritime adventurers from Spain and

Portugal who started to carve out the global system of inter-connected trade and finance that Pulitzer Prize-winning journalist Thomas L. Friedman calls 'Globalisation 1.0'. He sweepingly asserts that this was the globalisation of nations. Clearly, this was not entirely the case as, while European nation-states laid claim to overseas territory, much of the on-the-ground activity was carried out by privateers, or even by pirates, whose swaggering across the high seas was often, secretly, endorsed by European rulers.

What propelled the European geographical zeal from the 16th century onwards was not just economic avarice but the philosophical desire to evangelise others. Unlike the far older and more settled religions of eastern Asia, Europeans had come to adopt and adapt a relatively new Abrahamic religion called Christianity, which was still at the stage of forced conversions. It was just a few hundred years older than Islam.

On the death of the prophet Mohammed in 632 CE an evangelical zeal to proselytise also seized the Islamic religion. A century later, at the Battle of Tours, the combined European and largely Christian armies, under the Frankish leader Charles Martel, brought the geographical expansion-ist ambitions of the Umayyad caliphate to a halt. It was by all accounts a close-run thing. The caliphate's army withdrew into Iberia, now Spain and Portugal, where it was another seven centuries before Christianity was able to re-assert itself as the dominant religion. This was the age of Islam (al-Andalus) in the Iberian peninsula.

The evolving geographical push and shove between Christianity and Islam centred on the Mediterranean, and with it came an increasing interest in geography. It was here that maritime skills were honed on relatively benign waters, waters, it turned out, with favourable ocean currents around them.

The ability of some societies to utilise the sea for connectivity arose as technological progress improved both navigation and sea-worthiness. There are many disputed 'facts' when examining the maritime exploration of humanity, but it is through these journeys that much knowledge was gleaned to produce early maps.

The Book of Roads and Kingdoms (*Kitāb al Masālik w'al- Mamālik*) created by Persian geographer Ibn Khordabeh in 870 CE was constructed through interviews with traders, mostly Radhanite, plying goods between China and Europe. Later the *Tabula Rogeriana*, the work of the Islamic scholar Muhammad al-Idrisi, became the standard reference global map for a couple of centuries after the 12th century. Muhammad al-Idrisi was commissioned and aided in his project by the Norman Christian king, Roger II, whose domain was Palermo on the island of Sicily. Roger and al-Idrisi interviewed traders passing through their port city to glean more knowledge and experience. Their enterprise is clearly one of the most important early efforts at a cross-cultural and multilingual exploration of knowledge. The text that accompanies the maps is a work of 'economic geography', setting out the products and skills of places and people and the distances that separate them.

The Iberians had learned some vital things from 700 years of the Islamic caliphate, not least the Arabic numbering system and the usefulness of a zero. By the end of the

15th century such knowledge was being put to use enabling increasingly advanced ideas of navigation. Part of what propelled them (or made them believe drowning was not necessarily the end) was their Catholicism, almost as much as their desire for spices. The majority of South and Central America's citizens today are Catholic because that continent became the imperial stomping ground of Portugal and Spain. In contrast, the dominant religions of China did not suggest that through converting others to your beliefs so forcefully you could attain salvation.

The European Waldseemüller map of 1507, named after its creator Martin Waldseemüller, was the first geographic representation to note the existence of 'America'. Building on the voyages of Christopher Columbus in the 1490s, Italian merchant Amerigo Vespucci made several voyages out of Cadiz to South America, and it was Vespucci's reports that Waldseemüller was partly informed by and that he used in drawing the tentative boundaries for the Americas – and it was also the feminised version of the name Amerigo that lent itself to the naming of this new continent and peoples. In the 'natives' European geographers rediscovered a branch of humanity from which they had parted company some 40,000 years previously. Of course, they had no knowledge of that, just as today we will still lack knowledge of so much that will soon appear obvious.

In the 16th and 17th centuries European capital cities became rich on the plunder from the Americas and increasing trade with the East. These riches allowed new technologies to be developed and resources put aside for further exploration. Europeans travelled the world with the most advanced maritime technology yet seen, a growing grasp of navigation and an increasing store of maps to point the way. What was often most awaited on the quaysides of Cadiz, Lisbon and London were the new tastes and flavours that previously unknown foods brought.

Among the most valuable aspects of 16th-century global trade were the riches of biodiversity found elsewhere. The Americas revealed potatoes, sunflowers, cocoa, maize, tomatoes and blueberries. Lentils, chickpeas, onions and cucumbers came out of the Near East. Further into Asia were found chickens, rhubarb and millet. Africa yielded coffee, watermelons and oil palms, and later cocoa was planted there in huge quantities. The discoveries were bountiful and spurred an increasing volume of trade and a progressive re-ordering of the spatial distribution of plants and animals across the continents.

The chilli pepper, synonymous with the cuisine of the Indian subcontinent and Southeast Asia, was transported to India eastwards from Mexico only from the 16th century onwards, and it was introduced to new populations via the Portuguese colonial enclaves of Goa, Malacca and Macao. The oil palms that now extend across millions of hectares of Southeast Asia originated in Africa. The coffee that now so frequently fills our cups and senses may have first been traded from the Yemeni port of Mocha in the 15th century, but today countries as disparate as Brazil, Vietnam and Indonesia lead the world in coffee production.

To the Old World the American continents represented a huge potential resource. Plantations could be developed to turn energy from the sun into cotton and sugar, while energy from African slaves could be harnessed, and for many centuries Africa provided the muscle in much of the Americas. Yet the most valuable initial asset offered up by the Americas to its colonial invaders was the silver that came out of the mines of Mexico, Bolivia and Peru. This became the lifeblood of Iberia's colonial adventure. Millions died in pursuit of it, transporting it and defending it. In 1621 a Portuguese merchant, Gomes Solis, observed that silver 'wanders throughout all the world in its peregrinations before flocking to China, where it remains, as if at its natural centre'. This was a comment of great foresight, considering the process of globalisation that followed.

You might imagine that by now the Bolivian city of Potosí would be a place of prosperity, defined by fine buildings and pleasure gardens, given that it was here that so much of the wealth flowing out of the Americas originated. But it is not. Today Potosí is a city of 130,000 people and lies in the shadow of the towering Cerro Rica Mountain, 4,000 metres up in the Andean Altiplano. The mountain, an old volcanic dome that is rich in silver (106 grams per ton on average), has yielded up to 60,000 tons of the metal since the first Spanish excavations in 1545. Up on Cerro Rica, the 'silver mountain', miners still toil but for meagre rewards. In the 1620s this mountain gave up the ore to mint the silver *peso de ocho reales* (pieces of eight) – the world's first globally traded money – which funded wars, bought slaves, provided the liquidity for growing trade and was much coveted by pirates, of all persuasions. The pieces of eight minted in Potosí found their way across the world, and a disproportionate number of them made their way to

China from the 17th century onwards and helped oil the wheels of commerce there, as well as increasing envy from elsewhere at what China had to offer.

To extract the lifeblood that sustained Spanish colonialism, the conquistadores worked unknown thousands upon thousands of local natives to death in the enforced labour regime called *mita* before turning to Africa for slave labour. The Cerro Rica mines exist in an extremely hostile environment: the air is thin, the land is arid, and the conditions are extreme. Those caught up in *mita* servitude often spent months underground working 18-hour shifts. Human slaves were used to replace mules, which lasted mere months – slaves could last at least a year. What they produced became the most coveted thing on earth: the silver dollar. Where they produced it from became known as 'the mountain that eats men'.

What enabled power to arise out of trade was the medium through which trade was facilitated – the small, transportable objects of accepted value we all know as coins. Without coins, and later money, trade based around barter was limited by the idea of 'the co-coincidence of wants'. I might have a surplus of cattle, but do I really want your clay pots and do I want them now? In early societies this dilemma was overcome by the idea of imbuing a commodity that was 'in demand' with a specific tradable value – a shekel of barley, a koko of rice or even the life of a fellow human, a slave.

Initially, metals that were rare, beautiful and required a lot of energy to mine (often the energy of slave labour) became the resource upon which coins, and thus the idea of money, were based. The Zhou dynasty in China pioneered this approach from 800 BCE using bronze. The Spanish took this to another level with their silver dollars and gold doubloons in the 17th century.

Coinage and, later, promissory notes based around coinage enabled individuals and societies to use wealth in a transferable and tradable form. This meant that wealth or capital could be better stored and hoarded. It also meant that goods could be traded more freely and without direct barter – the tickets to enable the mutual satisfaction of co-coincidental wants became money itself, and money enabled flexibility and became power.

The Spanish discovery of vast silver mines in the Americas was perhaps the largest addition of 'new' monetary wealth created in human history. Silver was that 'tangible thing' that gave people confidence in its worth, whether it was used in the souks of Damascus, the trading hongs of Canton, the civilisations of Africa or the counting houses of Florence. Today, money is far more nebulous: it comes in many different forms, primarily binary digital code, out of which debts are born.

In 1602 the Dutch East Indies Company was established in what is now the Netherlands but in a region that was then called the United Provinces. The company was created to exploit the East; it had shareholders and has even been described as the first multinational company. Its model (and arch rival) was the English East India Company, which had been set up a year previously but was not initially as successful.

In the 17th century the United Provinces were the

dominant world power. The Dutch 'golden age' of painting was the height of European sophistication, and Dutchmen traversed the world: Adriaen Block determined that Manhattan was an island; Willem Schouten was the first to round Cape Horn; Abel Tasman was the first European to reach New Zealand, Tasmania and Fiji; and Williem Janszoon first saw the coast of Australia in 1606.

The Dutch were dominant before the English, but long enough ago to have been written out of much colonial history. The Netherlands is still associated with the tulip, the pretty flower that made its way into Europe from Turkey in the 16th century, and the tulip set in train what, it is often argued, was the world's first speculative bubble in 1637. In the taverns of Haarlem, collective stupidity gripped people desperate to be richer, pushing the price of some single bulbs up to 10 times the annual income of a skilled craftsman. Plants were big business in the colonial world.

Sir Joseph Banks (1743–1820) was one of the greatest of colonial plant collectors. His portrait adorns a wall in London's National Portrait Gallery, with his hand atop a pile of maps and a globe behind his arm. He was in many ways the most prominent geographer of his day – the late 18th and early 19th centuries, a time when colonial ambition was at its most vigorous. Banks's achievements are considerable, including the collection of the first botanical specimens from Australasia on Captain Cook's 1768–71 voyage of exploration. He was also the founder of the Royal Botanical Gardens at Kew. These achievements certainly rank alongside the peninsula, islands, 80 or so plants and Sydney suburb that bear his name. Banks's most significant discovery was the economic potential of the plants that he discovered, and it was to be a plant, *Papaver somniferum*, the opium poppy, that was to become a pivot on which the transition from the organic age to the fossil fuel age turned.

At the end of the 18th century the global balance of power had not yet tipped over to those European powers that had moved from an organic-based energy system to embracing fossil fuels and thereby setting in motion the industrial revolution. China was still, as it had been since the decline of the Roman Empire, the most sophisticated, complex and economically advanced nation-state in the world. However, the confidence of the Celestial Kingdom in its apparently unassailable dominance was not to last for long after European capitalists came visiting.

The diplomatic collision between Lord George Macartney, the representative of Britain's King George III, and the Chinese Emperor Quian Long in the Chinese capital of Beijing in 1793 has been much picked over by historians. The British came in pursuit of free trade and left with nothing more than a letter from the emperor to the king that stated: 'Our Celestial Empire possesses all things in prolific abundance and lacks no product within its own borders. There was therefore no need to import the manufactures of outside barbarians in exchange for our own produce.' To make the point about China's geographical integrity in such matters as trade and defence, Quian Long went on to say: 'Every inch of the territory of our empire is marked on the map, and the strictest vigilance is exercised over it all.'

As soon as Macartney left Beijing to rush back to Britain – where he arrived over a year later – the Chinese stepped up the defence of their sea ports. They had noted that in India the British, or rather their free-market proxy, the British East India Company, was progressing rapidly towards controlling a vast area of land simply through the expediency of promoting free trade, with the supposedly 'free' bit enforced by the barrel of a gun. At the time the British had the biggest collection of weaponry in the world. Yet, the Chinese did not know this. Had they done so, perhaps Emperor Quian Long's letter to George III might have been a little more conciliatory.

The constant struggle of European nations to have enough silver and gold to fuel their colonial ambitions had set them on a collision course with China, from whence came so much of what they wished to trade. The search for new colonies propelled Britain to the forefront of global power, and in the 18th century the British surged ahead of the Dutch, French, Spanish and Portuguese. With Adam Smith's *The Wealth of Nations* (1767–73) in one hand, the King James Bible in the other and a flotilla of warships never far over the horizon, the British set in train the rise of market capitalism that the United Provinces and before them the discovery of the Americas had spawned and out of which had sprung globalisation as we know it today. These became the primary economic forces of the new age, and this new age was the age of fossil fuel energy.

The British forced trade on China with weaponry, and what the Chinese were obliged to buy was opium. The weapons that the British took to China to enforce free trade were more and more often forged in furnaces powered by coal, dug from mines drained by steam engines and financed by people increasingly taught to have a colonial, geographical, ambition to expand a new empire of global proportions from Britain to every part of the world.

In 1800 Britain's mines were producing 13 million tons of coal; by 1900 this figure had risen to 225 million tons. This was the start of the short-lived fossil fuel epoch that has transformed the geography of the world. Fossil fuels have enabled humans to become better connected, and they have made it possible for the world's population to grow to over 7 billion people. In doing this, fossil fuels have provided us with the material and technological riches that our long-distant ancestors, mitochondrial Eve and her kith and kin, could never have imagined. They have also brought us to the edge of a potential environmental cataclysm.

2

GLOBALISATION

Only a few decades ago you could still find a lonely planet. Backpackers from the West carved out a well-worn trail through the magic of the East – Goa, Kathmandu, Bangkok and Bali. A few got off this well-beaten track, with some effort and often a Lonely Planet guidebook in their hand.

One place off the beaten track was Kagbeni on the southern edge of the near-mythical Nepali district of Lo Mustang. Here, north of the main ridgeline of the Himalayas, villages hang from precipitous, arid ridges perched above the wide flood plain of the Kali Gandaki River. Reaching this point a few decades ago required a six-day walk from the roadhead at Pokhara, and this involved climbing 3,000 metre high mountain passes, crossing makeshift bridges spanning raging, glacier-fed rivers and negotiating with villagers for whom electricity was still a distant aspiration.

To stand on the roof of the Red Lodge, then one of only two places offering lodgings in Kagbeni, after night had fallen, and before electricity had found its way up the valley, was to feel completely alone. The only illumination in the clear mountain air was the radiant light of the Milky Way in all its unpolluted clarity. You were disconnected from the rest of the world.

Today you can read a review of Kagbeni's Red Lodge on TripAdvisor. A mobile phone signal is intermittent, but that

Where is most remote? Greenland and Antarctica, and, of course,
Lonely Planet guidebooks exist for both destinations. Antarctica
is now described as 'this most arduous and demanding of holiday
destinations'. It would create at least 2.1 tonnes of CO_2 to fly from
London to Ushuaia at the very southern tip of Argentina, and there

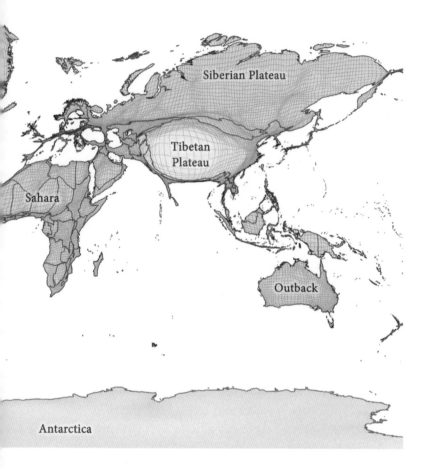

Siberian Plateau

Tibetan Plateau

Sahara

Outback

Antarctica

is still the small matter of 1,000 kilometres of the wildest ocean in the world to cover before you can make landfall in Antarctica at its most northern tip, the Chilean outpost called the General Bernardo O'Higgins base. It's best to avoid winter, but if it's remote you want, this is the place.

is probably no worse than in some parts of rural Britain. Thousands now pass through Kagbeni either full of trepidation and thinking of the long slog up to the Thorang La pass high above, which at 5,416 metres is the highest trekking pass in Nepal, or they are coming down and show all the signs of tired satisfaction at having made it over the pass. Kagbeni is now a significant stopping-off point on the famous Annapurna Circuit, Nepal's most popular trekking route.

In October 2014 a group of 40 tourists and guides did not make it over the Thorang La pass when, near to the top, they were hit by a snowstorm. The international TV crews took a couple of days to make it to this remote part of the Earth, but, like so many places that were once lonely and local, Kagbeni is now part of a globalised world. It is increasingly connected and today feels far less remote than it did just a few years ago.

Over the past 10,000 years humans lived their lives much as those who lived in Lo Mustang before the arrival of trekkers, the Internet and Gore-tex. Throughout almost all of human history, until extremely recently, the rest of the world was for most a closed book. Today a library of ideas, knowledge, peoples and technology stretches out into the remotest of communities. Our lives have become increasingly about other places. Home has become a whole planet, not simply the few square kilometres within our own horizon.

Such a monumental transformation has happened within just 200 years. You can grasp its scale by watching Hans Rosling's '200 countries, 200 years, 4 minutes'

YouTube video (see Further Exploration at the end of the book). Rosling, a Swedish medical doctor, demographer and statistician, is co-founder of Gapminder, a 'fact-based worldview' organisation, which has been influential in communicating a more positive perspective on global development than had previously often been the case. What Gapminder's informative video shows is that from a world of low life expectancy and material poverty, in just the course of a few generations we have (on average) become – everywhere – healthier and wealthier. Of course, this has not been an equitable or straightforward transformation, yet never has humanity been better off. Rosling is optimistic that such a trend will continue into the future and that the inequalities between nations that were set in train by colonialism will be further reduced very soon.

This chapter concerns those 200 years of fossil-fuel-led development and how the world today is more globalised – connected – than ever before. It is about how we have become less lonely and more global in our outlook. We may not all see things in as rose-tinted a perspective as Hans Rosling does, but then he is an optimist. It would be foolish to dismiss Rosling's enthusiasm for a better world, and the first step in achieving such a thing is believing that it can be so.

The forces that shaped global connectivity and propelled us technologically and economically so fast forward came into focus during the 19th century. The year 1842 remains one of the most significant moments in that journey – it was the year the First Opium War ended, and its reverberations are still felt today.

Reflecting on the First Opium War, a newly elected William Gladstone mused in the British House of Commons that 'a war more unjust in its origin, a war more calculated to cover this country with permanent disgrace I do not know'. However, such sentiments did not dissuade him from becoming colonial minister a few years later nor from serving as chancellor for the Opium War's lead instigator, Lord Palmerston, during the Second Opium War in the late 1850s. This was because, at heart, the Opium Wars were about business; they were bold assertions of liberal economics working in tandem with geo-political ambition, an early example of geo-political globalisation.

If you are a Chinese citizen reading this book you will need no introduction to the Opium Wars because they will be one of the core parts of your history curriculum. In Britain a similarly significant date might be 1066, 1815 or 1945. Nobody is taught about the Opium Wars in compulsory school history in Britain; everybody will be in China.

The year 1842 was the starting point of a 'century of humiliation' for the Chinese that lasted until Mao seized power in 1949 and instigated a revolution. It may appear strange that such a nationally negative appraisal of this period dominates the Chinese education system, but it enables this century-long downturn in Chinese fortunes to be placed at the feet of external, rather than internal, forces. Chinese school history does not dwell on internal cataclysms, such as the Great Leap Forward or the Cultural Revolution, nor even on the myopic worldview of the Qing dynasty at the start of the Opium Wars.

As numerous nations at various points in history have

concluded, it is easier to blame the outsider, the 'other'. Yet, the reasons for the commencement of the Opium Wars probably still baffle hundreds of millions of Chinese school children. Why would Britain, the most powerful nation in the world at the start of the 19th century, insist on trading in one of the world's most useful but pernicious drugs, opium? Stripped of any nuance or detail, the answer to this question is two-fold. First, Britain feared an almighty trade imbalance between Britain and China in the mid-19th century and the consequent downturn that this would create for British economic growth. Second, it was sheer unadulterated greed swept up by jingoistic racism and a mantra of free trade as a God-given right. The British thought free trade was something that imparted liberty to all that were touched by it. Such hubris led to Britain becoming the largest drug-dealer in the world, and the size of Britain's Victorian state-sponsored drug dealing would instil awe in today's cocaine barons.

When, in 1839, China, a sovereign nation, confiscated and destroyed 1,220 tons of opium that was flooding into the emperor's country, driving down life expectancy, productivity and social cohesion, the British response was to enforce its 'right' to trade. Merchants and free-marketers financed a military force provided by the British state, which assembled a mighty naval presence led by *Nemesis*, the first British iron-clad steam-powered warship. Collectively more firepower than had ever been known in China was brought together around the vessel the Chinese labelled the 'devil ship'. The British then asserted the will of free trade – the market – with brutality. Tens of thousands died and millions became addicted to opium. The Opium Wars touched every stratum and community in China.

Victory for Britain toppled China from its pre-eminent perch in the global power structure, a position it had held for nearly two millennia. Copious spoils of victory were extracted from the Chinese: not only some 800 tons of silver as full compensation for opium that had been destroyed, but also the costs of the military expedition, the island territory of Hong Kong as an addition to British territory, the removal of many trade tariffs for all goods traded with China and exemption from the Chinese legal system for British citizens. As business deals go, it was pretty good for the British. As a result, well into the 20th century Britain had a massive balance of trade surplus with China. The British could buy tea, spices, silks and porcelain at cheap prices with the silver they acquired from selling opium. Fortunes were made and many of those fortunes remain.

The Hong Kong and Shanghai Bank (later HSBC) was born out of this imperial relationship, and its initial wealth was made bankrolling British drug-dealers. This was a drugs market given legitimacy by state power. But in those days the opium poppy was just another plant.

Geography became an academic subject of increasing importance throughout the 19th century, and more often than not it was botany that travelled hand in hand with geography. In 1830, when the Royal Geographical Society (RGS) was formed in London, its initial meetings were held at the Royal Horticultural Society. Royal patronage gave geography, like botany before it, the status of a subject, a discipline, something that should be explored in seats of learning and not just used as a route map to colonial riches. The roots of the RGS were in the older Association for Promoting the Discovery of the Interior Parts of Africa (the African Association), another geographical endeavour of the uber-gardener Joseph Banks, this time dating from 1788.

The African Association confined its exploration to West Africa, where its most famous son, the Scottish explorer Mungo Park, 'discovered' the Niger River and later drowned in it. This event is 'memorialised' with the Mungo Park Medal, which is awarded for the furtherance of geographical knowledge in potentially hazardous environments. Nobody since Park has been awarded the medal posthumously. If the exploratory ambitions of these early African colonialists had drifted a few thousand kilometres south they might have stumbled on the mighty ruins of Great Zimbabwe, a sizeable African city that was the seat of power of an advanced state between the 11th and 14th centuries. However, it was not civilisations for which these African explorers were looking, but people, places and resources that could be exploited and also anything that might hinder such exploitation.

It has been argued that the lack of the tsetse fly in that part of Africa was a significant enabler of the Greater Zimbabwe civilisation. Whatever endowed this region with a developmental advantage, it was not until the early 20th century that a few British archaeologists came to accept that such monumental masonry could have been constructed by an African civilisation. In the post-colonial racist state of Rhodesia history was falsified until 1980 to suggest that the ruins of Greater Zimbabwe were the work of a non-African peoples – such is the scale and depth of the institutional racism that Africa has suffered at the hands of colonialists and post-colonialists. Many of these racists were geographers: racism is also, unfortunately, a very large part of the geographical tradition.

As the imperial imperative spread across Europe, huge swathes of the world were brought under the European economic and cultural yoke, and geography became the route map for this short period of global dominance. Successful assertions of power, such as that of Britain, framed geography into the 20th century, and early school geography books in Britain were underpinned with a social Darwinist pecking-order of racial superiority. That such books existed came about in part due to the work of Halford Mackinder, who became the first academic to teach geography in a British university, Oxford. He went on to set up nearby Reading University, run the London School of Economics, become an MP and claim to have climbed Mount Kenya with the help of a great many natives, some of whom he may have murdered.

Of course, as much as the British may wish to claim pre-eminence in the formulation of the academic discipline of geography, its roots, as we have seen, lie with Ptolemy, the myriad geographies of the Chinese state, Arabic astronomers and Indian mathematicians, and, more recently, with two great German geographers, Alexander van Humboldt (1769–1859) and Carl Ritter (1779–1859).

It remains relevant to observe that the first interest of Humboldt, and later Mackinder, was botany, just as with Joseph Banks. Where plants were, how they could be useful and how such biological parameters influenced human development were seen as key. Early geographers were also interested in exploration because so much was yet to be discovered – and in the world of botany so much remains undiscovered even today.

In the latter half of the 19th century, as industrialisation deepened and spread, technology progressed apace, and increasingly large energy resources were put to work to power all this progress, every aspect of which was often thought to be 'progressive'. European geographical and economic acquisitiveness turned its attention to Africa, especially African crops, mineral resources and souls to be saved.

The 'scramble for Africa' was as much a scramble for markets as for resources. Traditional societies were turned on their heads. At the Berlin conference of 1884–5, European leaders carved up Africa and Arabia with brevity and precision, drawing a series of straight lines that bore scant regard for topography or tribal and cultural affinity. Later, from Sudan to Iraq, conflicts would be rekindled along such hastily drawn lines.

While providing the modern industrial state with a cornucopia of potentially exploitable resources, geography also provided a growing body of alternative perspectives on how to live. Geographers tended to be conservative, and the subject of geography became the study of how to run colonies. But outside the academic discipline other ideas were taking shape, ideas that would later influence and remould geographic thinking and offer it the potential to overcome its militaristic and colonial past.

New world views were being formulated in tandem with the rise of the modern industrial state. Karl Marx and Friedrich Engels developed new theories of labour relations and the productive energy of capitalism. Alfred Russel Wallace and Charles Darwin realised that humanity lived within an evolutionary biosphere, and both were made Fellows of the Royal Geographical Society. Proto-geographical Victorian hippy Edward Carpenter challenged conventional ways of living from his home in rural Derbyshire, and the anarchist Peter Kropotkin expounded on radical politics while being an office-holder in the Russian Geographical Society. However, all these musings – although in some cases revolutionary – remained at the periphery of mainstream geography in their time.

The geography of grand Victorian buildings, of expeditions into uncharted territories and of a world map with a quarter of the land painted British red, coupled with a social Darwinian narrative suggesting the innate dominance of a particular people over others: these were the main perspectives of Victorian and Edwardian British geographers. And although the Royal Geographical Society, with its buildings in Kensington, is still the centre of British geography, things have moved on.

It is tempting to meander through the convulsions that litter the first half of the 20th century given that so much of great significance occurred: two world wars sandwiching a global economic crash, revolutions, revolts and new political and economic modes of thinking. Humanity realised that it had, within its own power, the ability to annihilate the world in a nuclear cataclysm. There was a retreat from empire, and new assertions of independence were made aplenty. This is a history that is familiar to many, and it forms the backbone of many schools' European history curriculums. There are still people alive today who bore witness to many of these events, although fewer and fewer. In May 2011, Claude Choules died at the age of 110. He was the last person alive who saw active service in the First World War, during which he witnessed the scuttling of the German Navy at Scapa Flow in 1919.

Whatever the social and political push and shove that shaped the first half of the century, one trend stands out as underpinning it all: our use of energy. In 1900 the world consumed 50 exajoules of energy. That's an enormous amount, with one exajoule being equal to one quintillion (10^{18}) joules, similar to the power released by an earthquake of magnitude 8.9 on the Richter scale. However, worldwide energy use had doubled by 1950, a rate of increase significantly greater than the rate of population growth over the same time period. Global per capita energy use rose faster still, but the geographical distribution of the increased energy consumption was uneven. Huge increases in energy usage within the USA propelled it to the forefront of global power, and the USA today requires as much energy in a year as the human world as a whole generated just 50 years ago. In contrast, until about 50 years ago there was barely any change in personal energy use in large parts of the world, particularly in China, in India and throughout most of Africa. Life expectancies in these underdeveloped parts of the world also remained persistently low: 1950 saw life expectancy reach only 42 in China and 33 in India. Child mortality was the key cause of such a low average in both countries.

For geographers today energy is the dynamic unifying everything, from plate tectonics and climate systems to the global economy and the culture of any place. We know that energy can neither be gained nor lost. Rather it moves from useful, ordered and concentrated energy to un-useful, disordered and dispersed energy. This is the principle of entropy, which has a complex definition but can broadly be expressed as the reality that everything, biological or not, will always eventually be heading towards breakdown: rot, rust and worms.

Every time we use energy, whether for growing food, for generating heat, for light or for motive power, we turn energy of high order into energy of low order. A lump of coal cannot be returned to a lump of coal once it has been burned. It becomes heat, light, gases and ash; it cannot be reformulated. This is something that even geographers half a century ago barely understood, and many today still do not fully grasp.

In the second half of the 20th century energy use increased exponentially, and that enabled a consequential explosive growth in human population. The billions of new, energy-rich, connected global citizens have become globalisation itself – they and how they are connected are its force, its being. More people can now live on remote mountain passes in Nepal, because those passes are now better connected. It takes great energy initially to connect everything to everything else in the human world as much as has been done today. Submarine cables, satellites, 3G and 4G networks all require more and more energy simply to sustain, let alone expand, them; roads require cars and lorries to travel on them, and this requires fuel.

Global energy use rose from 100 exajoules in 1950 to 500 exajoules by 2000. The world population also rose, from 2.5 billion in 1950 to 6 billion in 2000, but again not as rapidly as the growth in energy use. In addition, the energy mix changed over those 50 years, with oil and gas making a far greater contribution to the increasing demand for energy than coal. Oil was the energy of our new mobility.

We did become more energy efficient between 1950 and 2000, with energy intensity – the amount of energy required to create a unit of GDP (tons of oil per thousand dollars is the usual calculation for energy intensity) – falling in all regions of the world by at least half. Interestingly, most of this improvement in energy intensity has been in the last 20 years. Yet all the gains of such remarkable progress in energy efficiency have been offset by the absolute growth in consumption, especially the increase in consumption of the richest of people in the most affluent large countries.

Energy use propels globalisation, drives vast networks of connectivity and enables our planet to support a population far greater than Thomas Malthus might have imagined when he wrote the first version of his treatise on population limits at the end of the organic energy age, *Essay on the Principle of Population*, which was published in 1798. Today's extensive energy generation is necessary to maintain life as we know it, but it is not sufficient. We also require the organisational and institutional structures that enable that energy to be put to work for mass mutual benefit.

The United Nations was formed in 1945 to maintain international peace and security following the horror of the Second World War, with its unprecedented number of casualties, nuclear denouement and the substantial impoverishment it inflicted upon all participant nations bar the ultimate victor, the United States. It was the USA's dominance over global energy supplies that made it almost certain that it would prevail in a war at that time. The nuclear bombs it unleashed on the already inevitably defeated Japanese nation in 1945 were the ultimate expression of that power. As Robert Oppenheimer, one of the architects of nuclear bombs, said, quoting the Hindu holy text the *Bhagavadgitā*: 'Now I am become death, the destroyer of worlds.'

Interest in global governance and international oversight was further strengthened in the light of the global flowering of democracy. Colonial nations like Britain divested themselves of their empires (they often had little choice), while labour rights and welfarism began to be developed by left-leaning governments, and the USA finally started to focus on its racial inequalities. All over the world people were becoming more equal, and towards the end of the 20th century equality was spreading into the hitherto uncharted territories of sexual equalities, meaningful gender equality and racial equality.

There remains much distance to travel in moves towards addressing the uneven distribution of equality, but, like Hans Rosling's graphs of health and income, a graph plotting social and economic equalities would see it moving over the past 200 years from the unequal, prejudiced and racist corner of the chart towards the more equal, less prejudiced corner, at least through to the early 1980s. Whether this trend continues is very much unknown at present, because neo-liberal forces of inequality have significantly re-asserted themselves in some countries in recent decades.

When it comes to the movement of people, to transport, to communication technology and to finance and trade, more has changed in the world in the seven decades since the end of the Second World War than in the seven centuries that preceded that war. This is why it is commonly perceived that globalisation is a key phenomenon of our recent times, part of what defines the geographic transformation of humanity that is now fully under way, now that we can talk through wires, send pictures through the air – and now we can fly.

Today air travel is ubiquitous, and it is estimated that there will be 8 billion passenger journeys a year by 2030. Airfreight is now an affordable option for moving goods, particularly perishable items, at least for the richer consumers in rich countries, who can buy supermarket cut flowers or strawberries for the Christmas Day dinner table in winter. However, most of the stuff we consume comes from afar in metal boxes ('one twenty-foot long unit') and travels far more slowly by ship, not through the air.

The development of the shipping container in the 1950s had a significant impact on global trade by enabling economies of scale to take hold and thereby decreasing the unit cost of transporting goods. Today products, even relatively inexpensive goods, can be moved across the world and the trade is still profitable. The change is all about geographical scale. Our opening account of the maiden voyage of the

MSC *Oscar* is illustrative of that application of energy and power to create transport geographies on a scale never before imagined possible.

The shipping container can tell us many things about the state of the global economy. For example, the relative trade imbalances between the USA and China have produced a massive build-up of shipping containers in the USA. The global shipping line Maersk spends $1 billion a year moving empty containers around the world, back to where they can be filled up. As much as 82 per cent of shipping containers in use worldwide were manufactured in China because that is where they are predominantly being filled, at present.

Unsurprisingly for an object that spends 56 per cent of its lifetime sitting idle and empty in some port stack, innovative ideas for utilising containers heading back east from Europe have been tried, including filling them with the waste that the accelerated consumption they facilitate generates. Every year the UK sends 1.7 million tons of waste to China for recycling, much of it in shipping containers. This new trade in rubbish has made great fortunes, including the $4 or $5 billion accumulated by China's wealthiest self-made woman, Zhang Yin, whose company, Nine Dragons Paper, imports waste paper from the USA to China for recycling.

What has smoothed the way for the massive surge in global trade in the period following the Second World War isn't just technology, greater energy efficiency or increasing global wealth. The formulation of a constantly evolving set of global trade rules has been crucial. Sometimes these rule agreements are bi-lateral, between two distinct countries, but predominantly they are multi-lateral, agreements between multiple nations.

Over three summer weeks in 1944 the foundations of the present architecture of the global trade system were constructed at the Bretton Woods Conference by representatives of the victorious Allied nations just before the end of the Second World War. Bretton Woods was a sleepy – but luxurious – mountain resort in New Hampshire, USA. From Bretton Woods onwards there has been an on-going aspiration in the world of orthodox economics for stable exchange rates, mobile capital and minimal tariff barriers to facilitate smoother commercial exchange in order to maximise the flows of global trade and, so it is posited, for global prosperity to increase as fast as it possibly can. The complexity of the inter-connected relationships between nation-states and private capital has led to a series of torturous international negotiations, mostly on-going.

Openness to trade has its critics. Some argue that global trade rules are loaded to the advantage of the countries that are already most successful on the global trading stage. Others point to the asymmetrical impact of these rules when agricultural produce remains stoutly defended from competition in many of the most affluent countries, like Japan and the USA, but manufactured goods move with minimal restriction between countries. Many bemoan the loss of national sovereignty to the titans of global trade: transnational corporations.

Today a new global trade agreement is being drawn up to try to regulate all our commerce, and it is being resisted by increasing numbers of people who see it as a charter to allow despots to rule. The Transatlantic Trade and Investment Partnership (TTIP) sounds benign but very likely isn't. It will tie up the USA and EU into a single market with very little local protection. It will permit global corporations to further side-step due legal process in any signatory nations in which they operate.

Emboldened by existing skewed regulation, tobacco manufacturer Philip Morris is bringing legal action against Australia and Uruguay to demand either compensation for, or the repeal of, public health legislation that is aimed at reducing the harm that cigarettes cause. In Europe Swedish energy multinational Vattenfall is suing the German government for $6 billion through the Washington-based International Centre for Settlement of Investment Disputes (ICSID), an offshoot of the World Bank. This is as a consequence of Germany's policy of phasing out nuclear power. Vattenfall has majority holdings in two redundant German nuclear reactors. Nor will Vattenfall shoulder the full decommissioning cost of these reactors – a cost that will be in perpetuity, given the nature of very slowly decaying nuclear radioactivity.

The new legal imperialism may not involve gunboats, opium or colonial belligerence, but the echoes of Britain's enforcement of the free trade in opium in the mid-19th century are beginning to be heard again. Governments will be unable to set social and economic priorities according to the democratic will of their populations without paying compensation for potential lost business profits to private (often foreign) companies operating in the markets that are being legislated into submission. This free-trade template, which was tempered through the growth of democracy and equality throughout most of the 20th century, is re-asserting itself across the globe with increasing vigour today.

Although goods cannot be moved from place to place without it taking some time and much energy, few such geographical barriers remain when it comes to capital, which has gone digital. With the rise of offshore financial centres, capital mobility is now beyond transparent regulation. Globalisation has resulted in the emergence of a financial pirate class, which can pounce anywhere, strip wealth from a place and its people and disappear into a financial magic box where no prying eyes may penetrate. Most such jurisdictions that provide tax haven anonymity operate under the Union Jack and are Crown dependencies of the United Kingdom, from the Cayman Islands to the Bahamas. Even the 'City of London', with its separate legal jurisdiction, operates much like an offshore economy, just one that happens to be in the heart of the UK's capital city.

The digital world has enabled the world of capital to develop an alchemistic relationship with money. Banks and other financial institutions employ the sleight of hand that is fractional reserve banking or, in simple terms, lending money they don't have. A consequence of this financial merry-go-round is the creation of a massive transactional, largely unproductive economy – the shadow economy – that is both geographically rootless and beholden to no nation-state.

The 'shadow' economy that has been created by hypermobile capital has created market distortions and criminality on an industrial scale. In 2011 the British-based Tax

Justice Network published research that revealed that globally $3.132 trillion of revenue a year was being lost to governments through tax evasion (and this is almost certainly a significant underestimate). That is approximately 5 per cent of global GDP. The Tax Justice Network estimates that $1 in every $6 is hidden from the tax authorities worldwide. Furthermore, the ability easily to side-step regulatory structures has facilitated a massive rise in criminal fraud, often aided and abetted by our increasing digital presence.

'Old school' criminal notions of having to launder money to 'clean' dirty, criminal money are largely gone, because so much of the money that circulates is 'dirty' and so much is unaccountable. As has already been pointed out, often complicit in this criminality enabled by globalisation are those same banks and financial institutions that project an image of being sturdy, honest and upright businesses. Often this is a marketing illusion. The culture that underpins the acceptability of tax avoidance goes right to the top of political elites. In January 2014, British prime minister David Cameron said in a speech to the Federation of Small Businesses conference, 'frankly I don't like any taxes'.

Much that is currently endorsed today as fair may perhaps, in the future, be seen as despotic. Since the Opium Wars of the 19th century, Britain has remained at the heart of the global drugs trade, but this time it is financial institutions and lax regulation that facilitate such activity. UK bank HSBC was fined $1.9 billion in 2012 by the US authorities for money laundering, and the US government described HSBC as a conduit 'for drug kingpins and rogue nations'. HSBC, a global institution though nominally British, is fuelled by the fossil fuel economy, for its global reach, the trade it facilitates (legal or otherwise) and the profits it accrues. It remains a totemic institution of globalisation, right from the many ways its roots are entwined in imperialism, up through all its shoots and branches.

This is where globalisation has arrived. We live in a world that has never been more connected – a result of humankind's ability to harness so much energy to its will. We also live in a world of immense inequality, with the mass of global poor holding but a minute fraction of the wealth of the relatively few rich. How we have got to this place at this time is a result of the fossil fuel revolution. It has left humanity – all of humanity, rich and poor – with an imminent environmental crisis looming on an unprecedented global scale. Not one of the regional environmental and economic collapses that have littered human history can compare to the future problems we face, and no amount of wealth can insulate any group of people from this reality, wherever they live. This is how globalised we are today.

Many theories abound around globalisation. These ideas are not held in isolation but are propelled around an increasingly connected global population, through tweets, texts, blogs and posts. Sometimes people even call each other and talk. Some glibly speak of 'McDonaldisation', where we all are subsumed by a single global culture – apparently involving fries and a shake. Others emphasise the huge increase in global trade, the rise of China (the re-assertion of China would be more accurate) and a global capital market that has no national identity and no focus beyond 'the bottom line'.

In the past the discipline of geography was part of the problem – it was the subject that promoted colonialism – but today the geographical perspective is critically to consider many different angles and worldviews. This is because the geographical perspective acknowledges that all of these global interactions – the trade, the tweets, the holidays and the national rivalries – are going to be driven forwards by what energy we can harness and what consequences will follow from such energy use. This connects us all. This is at the core of globalisation. We live on a lonely planet no more, but one that is being split asunder in many ways.

3

EQUALITY

We live in an unequal world. For most of human life on earth we were remarkably equal as regards our relative power and (few) possessions – that is, until the advent of agriculture. Today, in terms of the differences in the level of resources available to people in different parts of the planet, we have never been as unequal as we are now.

The patterns of economic inequality that frame our world are complex. On one level there is a strong argument to be made that we are becoming economically more equal, with hundreds of millions being lifted out of poverty, particularly in China and India – that is, as long as you deduct from your calculations the increasing wealth of the world's richest 1 per cent. Within that global 1 per cent the scale of inequality is huge. Those at the very apex of global wealth (the 0.01 per cent) hold a huge and ever-increasing store of wealth. Economic analysis of wealth formation and distribution strongly suggests that the 'trickle-up' tendency for wealth has both accelerated in recent years and is entrenched by our current globalised political and economic system.

A stark example of the current extremes of inequality is found within Mumbai, India. Some 12 kilometres separate Dharavi, once the largest slum in the world with up to 500,000 occupants, and Antilia – the world's first £1 billion home. Industrialist Mukesh Ambani, whose family spreads out through Antilia's 27 floors, is India's richest, and the world's 39th richest, individual, and he is clearly in possession of a sense of irony, because near to him at Antilia is the rather more modest home of Indian social activist Mahatma Gandhi, who sometimes lived there between 1917 and 1934.

It is easy to become despondent when you grasp the full scale of current economic inequalities. However, there are other ways in which equality is rapidly increasing, and it is a huge mistake to ignore those. Some of these improvements involve things that are much more important than money – above all health.

We alive today are the longest lived humans who have ever lived. We eat more regularly and better than ever before, and we know so much more about medicine and how to live healthily than any previous generation ever did. Furthermore, such advances have not been confined to a few parts of the world. Almost no area of the world is untouched by improvements in health. There are pandemics that interrupt progress, such as influenza in 1918–19 and AIDS, especially in the 1980s and 1990s, but over the last two centuries health has improved almost immeasurably, especially for the young. Never before have so few parents experienced the death of a child. Indeed, many times fewer people suffer such experiences in comparison to the tragedies that so many of their great-grandparents endured – and all this despite great population growth.

In 2013 life expectancy in the UK was 81 years and in China it was 77 years. The greatest proportional increase seen in recent years has been in India, where life expectancy now stands at 66 years. All of this has happened alongside the greatest surge in human population the world has ever experienced – from 1.3 billion in 1865 to 7.3 billion by 2015. Perhaps most startling of all is that even as population has surged in a 'few favoured generations' (to quote Charles Darwin), the amount of nutrition available to humanity has increased by an even greater amount. Darwin was not talking of people but other species that experienced population booms when he mentioned 'favoured generations'. He had no way of knowing that he was writing at the very beginnings of just such a boom for his own species.

According to the World Heath Organisation, between 1985 and 2015 the amount of available global nutrition has grown by 65 per cent in total. The average per person per day calorie intake across the globe has increased from 2,655 calories to 2,940 calories. All these changes occurred at the same time as 2.5 billion people were added to the global population. The Global Nutrition Report of 2014 concluded: 'With regard to food supply, as undernourishment declines, over-acquisition of calories is rising. This means that the share of the population that has a healthy food supply – neither undernourished nor experiencing over acquisition – remains constant instead of increasing.'

Of course, these developments in health and longevity have been uneven across both time and place. Events like the HIV/AIDS pandemic did have and continue to have a major impact on life expectancy, particularly in sub-Saharan African nations. The flu pandemic of 1918–19 temporarily but significantly reduced life expectancy in the many nations it touched.

Although it is repetitive to keep mentioning it, it really is worth saying again: you should take a look, if you have not already, at the visual impact of Hans Rosling's interactive graphs on his Gapminder website. For many it is revelatory. It is Rosling's aim to fight devastating ignorance with a fact-based worldview that everyone can understand. The long-term trend he illustrates between life expectancy and wealth from 1800 is obvious: we are all moving towards the healthier and – to a lesser degree – the wealthier corner of the graph. Rosling provides the evidence to back up the claim that we live 'in the best of times'. Just as there are remarkably bad events still happening, there is also much remarkable good news, and a great deal of it concerns equality.

It was Hans Rosling who first explained to the wider public that the world had passed the 'peak baby' point – where the absolute number of births peaks and starts to decline – in 1990. The birth rate peak (the number of babies per woman) had been reached more than two decades earlier even than that in 1971. As fewer and fewer babies are born, those there are can be better cared for, and they are more likely both to be inoculated and to sleep under mosquito nets. They are also more likely to be better fed and have a greater likelihood of being cared for by parents who themselves are living more fulfilled and less desperate lives, which ideally enables a greater degree of care both by them and, later, of them.

Since AIDS deaths began to abate, global health inequalities have narrowed. This should be a huge source of celebration, despite the fact that there is still so much that can be achieved. It is only because of these improvements that we expect to reach a population of 10 billion people by the year 2100, if not before. Were it not for these health improvements, families would likely have far more children and the population would rise a long way above 10 billion before it stabilised. However, even if we are adequately to feed the 10 billion global population, we will require the production of yet more food, coupled with a dramatic reduction in food waste and over-eating.

The major reason for the growing equality in health, rapidly reduced fertility rates and increasing life expectancies is the revolution in the position of women. The feminist movement has been the most effective force for greater equality ever seen in human history. It is mostly only within the last century that women have made significant progress towards being treated equally with men. Contrary to popular misconceptions, much of the most significant current progress has been made within some of the world's poorest and most underdeveloped countries, such as Rwanda, Nicaragua and Burundi – all of which stand above the United Kingdom and the United States in the World Economic Forum's Gender Gap Index.

It was the progressive – and now very long-lived – Scandinavians who were the first to legislate towards achieving equal political rights for men and women regardless of economic status (Norway in 1913 and two years later in Denmark). Today the only two 'democracies' in the world that do not allow women to vote are Saudi Arabia and the Vatican – and we use the word democracy extremely loosely in reference to these two countries. Women's 'herstory' is one of subordination and subjugation in respect to men. From the Athenian and Roman Empires, where goddesses existed but no political representation was afforded to women, through to the later spread of the major Abrahamic religions (Judaism, Christianity and Islam) with their very fixed views on how men and women should behave and be treated, women have been remorselessly discriminated against, and this has been argued to be the natural way of things.

Today there is still a great distance to be travelled in achieving gender equality, even in the progressive nations of Scandinavia, but particularly in the countries of Yemen, Pakistan and Chad, which are adjudged by the 2013 Global Gender Gap Report to be the most gender unequal nations in the world. But everywhere the gaps are narrowing.

Never has there been less gender discrimination globally, although progress in this area is both unevenly distributed geographically and unbalanced in respect to the 'pillars' of discrimination. For example, the latest Global Gender Gap Report demonstrates quite high and improving levels of equality in health and survival and in educational attainment, but with substantial ground to be made up in the area of economic participation and opportunity. However, it is in the area of political participation – power and decision-making – that women are still the most substantially unequal.

As we write, Brazil, Bangladesh and Germany are the largest nations with a female head of state. Currently, 600 million people worldwide have a female national political leader, and many other nations have previously had a female political leader. Yet at the local, national and international scale, within business, the judiciary and in terms of political representation, women remain significantly under-represented. Although no one should be complacent about the challenges that still need to be met in order to achieve full gender equality globally, given the progress made from just over a century ago, there is reason for optimism for future progress. What could, should and almost certainly will be achieved in the next century should just about seal the deal: men – women – equal rights everywhere.

Human rights for individuals across the range of human sexuality have expanded, although again in a highly uneven manner geographically at the global and to some extent national level. That a Conservative government in the UK should endorse and legislate for same-sex marriage would have been difficult to imagine even as recently as the 'swinging sixties', or even in the 1990s. Through the first decade and a half of the 21st century 11 European nations introduced legislation permitting same-sex marriage, and in Ireland a majority of the electorate voted for it. The possibility has been discussed in India and China and is widely accepted across the USA, with over 50 per cent of the population supporting it. In June 2015 legislation enabling same-sex marriage was enacted in the United States, making it the 21st nation-state to achieve this by that point. Elsewhere the situation is not so positive.

Clearly Gay Pride marches in cities as diverse as Kampala, Uganda, and Tehran, Iran, would be high-risk activities and unimaginable at present. The persecution experienced by many – in countries such as Iran and Uganda – concerning their sexuality remains shocking; in some cases fatal. Yet 'the love that dare not speak its name' is now heard, seen and increasingly protected by numerous national legal frameworks, and it is hoped that these newly won freedoms will progress in the 21st century to a point in the future when to be gay (and much else that was until recently proscribed) will be unremarkable.

Yet regardless of the many positive stories of rapidly increasing equality, globally it is also the case that the resources that enrich our lives are not distributed equally, and this has had a significant influence on the formation, retention and spread of wealth, both historically and spatially. Humans remain most unequal not by gender, race or sexuality, but by wealth. Ironically, it is the wealthiest nations on earth that have most denuded their natural resources – but wealth is about power far more than it is about resources to hand. However, resources matter and will soon matter more.

Becoming more economically and socially equal in the future is one of the greatest geographical challenges we face as a species. Land is not equally fertile, and available energy sources – organic and fossil, wind and water – are distributed unequally. Even incoming solar radiation is unevenly spread across the Earth's surface. Fresh water resources, fish stocks and biodiversity all exhibit significant spatial variability.

Take energy. Exploitable sunlight varies across the Earth's surface. It is potentially at its highest across the Sahara Desert, the Namib Desert and a northern swathe of the Great Australian Desert. Wind energy potential is greatest in areas such as the west coast of the British Isles, Argentina, Greenland and the prairie states of the United States.

Hydro-electricity gives the best returns in mountainous areas with low tectonic risk factors, where it is often already being exploited, such as in the Snowy Mountain project in Australia.

Tectonic concerns have not dampened China's ardour for hydro-electricity. Of the six dams worldwide with an output in excess of 10GW, three are in China, and this includes the largest of them all, the Three Gorges Dam, which boasts an energy output of around 22GW; for comparison, 1GW would be the output of a large coal-powered power plant, and total UK energy demand averaged 36GW and peaked at 57.5GW in 2012. All these Chinese mega-dams are built in areas affected by significant earthquakes, and building them required a huge displacement of people. Other forms of raising more power are potentially less disruptive to other humans in the long term.

When it comes to the possibilities for tidal power, the British Isles are incredibly well endowed. They not only have the greatest global potential for tidal energy, but the coastal topography to exploit it with a high degree of effect-iveness. Coastal northeast Canada and the Pacific coast of Colombia also have a great deal of tidal energy potential.

A tidal barrage across the River Severn could generate up to 7 per cent of the UK's electricity demand and, although expensive, would cost only around half of the equivalent in terms of nuclear power stations producing a similar amount of energy. A tidal barrage would, however, also have sig-nificant impacts on the estuarine ecosystems in the Severn estuary as an act of significant intervention in coastal mor-phology (influencing coastal processes and features). This would have negative impacts on factors as wide-ranging as wading bird populations and access to port facilities.

Then there are fossil fuels. The inequitable distribution of fossil fuel reserves, coupled with differences in the ease of their extraction and thus their price, has shaped our world more in the past few decades than any other geological factor. Many wars are wars for oil. Modern-day imperialism is more often the search for more black gold rather than the older search for more markets.

The United States would not have been able to propel itself to the forefront of global power without its ability to tap into abundant and easily accessible fossil fuels. From 1900 to 1971, when the US reached 'peak oil', the point of its highest level of production, it was the largest global producer of oil in the world, and in the 1950s the US was supplying over 80 per cent of all oil consumed globally. In the oil age this simple fact explains the magnitude of the USA's economic, political and military power.

Geographically, the US has been fortunate in many ways, but the plentiful, cheap and seemingly inexhaustible energy resources that it had, or appeared to have, in perpetuity were the fundamental driver in its ascendancy to global economic dominance. Likewise, without an abundance of coal Britain would have lacked the energy reserves to drive the industrial revolution, the building of an empire and the accumulation of wealth over many subsequent generations. Place really did and does matter. Other parts of the world also had coal, but were not so well situated between West Africa and the Americas to use ocean currents and wind to facilitate the new trade in slaves, goods and manufacturing that the British also developed.

While it is undoubtedly the case that humanity has progressed as a consequence of cooperation, conviviality and even altruism, such motivations have always been set against what some have coined 'the animal spirits'. Competition for scarce and unequally distributed resources has been a central human dilemma. Treating others poorly, including as goods to be traded, has also been part of greed, which has driven humanity to scheme, speculate and hoard. Indeed, it has led humans often to define their individual sense of themselves in respect to the accumulation of resources. Once what appeared to many to matter most was cattle or seashells; now it might be cars or real estate.

Today there are plenty of people, perhaps even most people, who feel that human inequality is a natural state of being, just as there were once people who believed slavery was natural and women were naturally inferior. We are all born with different talents, some of which are more desired or easier to sell or have greater utility at one time or in one place. Be born female almost anywhere a century ago and you were born subservient.

In reality for the majority of the world's population where, when and why you were born are more important indicators of your 'position', or social and economic status, than any allusion to merit. Knowledge is unevenly distributed (and often simply 'wrong'). The resources to actuate development are unevenly distributed. Your life opportunities today, almost entirely regardless of merit, are substantially determined by where you were born and to whom, both on the local and global scale.

Be born a boy to a billionaire in the US and it would be hard not to receive what many might call a 'privileged education' and to later 'do well'. Be born a girl in the worst street of a slum in Brazil and almost no matter how amazing you are, you will have done well if you manage to survive and reproduce. And that may be a much greater achievement than that of the billionaire's son when measured against what you are likely to achieve.

There are exceptions to the generally robust predictability of outcomes from circumstance; an example is the story of William Kamkwamba from Malawi. Kamkwamba's is the story of a young man who was forced out of education by poverty. Yet through the knowledge gained from reading library books, he used scrap materials to build a series of wind-powered generators, which brought electricity to his remote village, and he went on to study environmental science and engineering at a US Ivy League university, Dartmouth College. How many Williams remain unfulfilled because of inequities in our access to education? The evidence is positive, suggesting fewer and fewer lack access to education, and this may mean more and more competition for the old European and US elite universities.

In India (and still in Britain it has to be said) much of the education system has remained stuck in the colonial mind-set bequeathed by the British – it remains a system built down from an elite apex. However, a significant challenge to this is being made by India's third richest man, Azim Premji, an IT entrepreneur, who, as founder and CEO of the company Wipro, has amassed a personal fortune of around $16 billion. Premji has already given away 25 per cent of that wealth, primarily to set up a trust to fund a new university in Bengaluru. This university aims to train teachers to go out into India's rural heartlands to transform educational standards for the poor. This is, Premji believes, the key to improvement, although he also says that such giving doesn't 'continue to substitute the responsibility of government'. The untapped educational talent pool in India has the ability to transform that country only if improved governance and greater business and political transparency were to make that possible.

The United Nations Human Development Index (HDI) – the world's acknowledged 'league table' of national economic and social well-being and our common measure of global inequality between nation-states – is a remarkably simple index. It is one that any mildly mathematically literate student could have thought up. It simply averages life expectancy, GDP per capita and educational attainment in a very unsophisticated way, with the rationale being that if GDP were not included it would not be taken seriously by economists. At the beginning of 2015 Norway was at the top and Niger at the bottom of the Index. The chasm between them must seem intractable to people in both nations. Yet there will be poor people in Norway and rich people in Niger. Inequality is always a relative measure. It is about disparity.

So what forces have conspired to make Niger the nation with the lowest level of well-being on earth? Certainly governance in Niger has not been to Norwegian standards.

Since its independence from France in 1958 Niger has been through a cycle of four coups d'état and is now on its seventh republic under a new and democratic constitution that came into force in only 2010.

Environmentally today Niger has significantly constrained capability as a consequence of its limited agricultural (bio-capacity) potential. Although it encompasses 3.12 million square kilometres – making it about twice the size of France and three times larger than Norway – 81 per cent of the country is arid desert with no agricultural potential. Of the remaining land – at the desert margins – 17.2 per cent is considered to be extremely or very vulnerable to desertification by the Soil Division of the US Natural Resources Conservation Service. The remaining 1.8 per cent of land that is deemed productive, and not at present vulnerable to desertification, must support a population of 16 million people. In Europe Slovenia is similar in size to Niger's productive land, and Slovenia supports a population of only 2 million. However, Norway, too, has little agriculturally productive land.

Niger is rich in energy resources, with massive solar energy potential and huge uranium reserves, the exploitation of which accounts for 75 per cent of the value of the nation's exports. This uranium is predominantly used to power the nuclear power stations of France, Niger's old colonial master. By 2014 Niger's uranium generated three-quarters of France's electricity production. Niger is a nuclear state without nuclear power, and knocking at its southern door is the Islamist terror group Boko Haram.

One translation of Boko Haram is 'Western influence is a sin', which – given much of the impact of Western influence in Africa to date – is understandable.

Although climate change predictions for Niger are not as extreme as in other places, it matters little because in such a marginal environment small changes can have big impacts. Niger will get drier, and its capability to feed itself will decline. Across the Sahel, the southern Saharan desert margins, drought and famine have often stalked. As recently as 2012 famine had to be fended off with food aid, and in 2006 and 2010 similar crises occurred. The economist Amartya Sen believed that famines do not occur in functioning democracies because their leaders must be more responsive to the demands of the citizens. Sen has written widely and to great acclaim on the causes of famines, contributing towards the award of a Nobel Prize in 1998, and he also designed the Human Development Index with Niger and Norway at such extremes.

Although it is true that Niger has not yet enjoyed a functioning democracy and the good governance that would enable it to address food shortages, it is also the case that, even if such positive governance prevailed, it would be fundamentally constrained by a growing population, which is subsisting on a declining agricultural base. The reality is that Niger has insufficient capability – food, energy and water – to redistribute even if an effective mechanism (governance) to engage in such a task could be created. In the medium term Niger may become a permanent recipient of food aid from countries like Norway, and the probability of its making much headway in the UN's HDI league table seems remote. Such inequalities are predominantly geographical, but not wholly.

Even when blessed with a resource that may allow Niger 'to pay its way in the world' – uranium – the French state-owned company, Areva, that exploits this resource has impoverished the Niger state by negotiating tax exemptions. In 2010 these were worth €320 million, and Areva contributed only €459 million in tax payments – that is, just 13 per cent of the €3.5 billion total market value of the uranium mined by Areva in Niger in that year. This might appear a reasonable level of taxation, but bear in mind that the UK levies a 50 per cent rate of petroleum revenue tax and then charges on top of that corporation tax on company profits, until recently at a rate of 20 per cent, although it had been 28 per cent in 2010. As the UK tries to 'win the global race' and rekindle what some see as its former colonial era glories, its government slashes corporation tax in a race to the tax haven bottom. Former centres of empire can find life after colonialisation difficult to adjust to just as much as former colonies do.

Colonialism's political legacy has a persistently long tail. Geographical inequalities can be amplified by political and economic inequalities. In Niger's case a nexus of geographical, economic, historical and political factors leaves it the most underdeveloped part of our world. It is, as geographers are apt to say, on the periphery. Niger is on the edge of everywhere.

Of course, there was a time when Niger was the centre of everywhere, when traders would arrive from nearby

Timbuktu, when Lake Chad was not so depleted as it now is but was a great water source, when the River Niger was a major highway of trade.

The capital of Norway, Oslo, lies almost directly north of Niger's capital Niamey. About 5,150 kilometres and a world of difference separate them. A population of 5 million Norwegians are ranked to be living the high-life by global standards on those material quality of life indices, and because Norway is an equitable country, most of them actually get to enjoy this good life. They have also been able to accumulate wealth from the export of their own hydro-carbon geological resources to the extent that they currently hold, collectively, about 1.3 per cent of all quoted equity (shares, fixed assets, debt bonds) found anywhere on the planet. Norway is among the most equal of nations in the world in almost every sphere of life, economically, sexually, politically and socially. It is also the fifth least corrupt country in the world.

It might be tempting to attribute Norway's developmental success to something in Nordic genes or perhaps their culture and then hold that up against Niger and say, 'there you go – they were just born better'. Such a perspective would dismiss the thousands of years of history that have led us to the formation of these geographical inequalities. It would also neglect the role of the ability to capture energy in shaping development. An even more obvious critique of such an argument is that the Norwegians were dirt poor just a couple of centuries ago; just as Niger is now and was not in the past. In both cases the peoples' genes are almost identical to before.

What links the development of different places is the

terms of the trade between them: trade in raw materials, energy, produce, manufactured goods, services, technology, knowledge and even philosophies. In 2012 Niger imported $1.04 million of goods from Norway but exported nothing directly to that country. What connects these two national extremes of development is not trade but aid. Norway contributed $7.7 million in aid directly to Niger in 2013. Of course, Niger did export uranium to France, and France did export goods to Norway, and so there are indirect trade connections, but whichever way you look at it, aid is now the key economic relationship between Norway and Niger.

Today the availability of immediate resources within a nation-state's borders or sea area is less pertinent because trade, mobile capital and the historical accumulation of wealth in specific places frame exactly who has access to resources. Huge wealth may reside in places that were also once on the edge of the global economy, the periphery. Dubai thrusts its economic assertiveness skywards. Luanda, Angola's capital city, has the reputation of being, at present, one of the most expensive cities on earth. In the depths of the Amazonian rainforest in Brazil lies Manaus, a city of 2 million people, which thrives with a significant degree of prosperity, despite having only two road connections, an airport and 1,500 kilometres of river connecting it to the outside world. Manaus used to be called Cidade da Barra do Rio Negro (the City of the Margins of the Black River). Remote these places may be, but they are also prosperous.

The expansion of colonies into sparsely populated parts of the world afforded European nations a demographic pressure valve. Places to expand into included Australia or Canada or even Manaus, when the rubber boom of the late 19th century brought thousands flocking to make their fortunes, leaving the 'pinched' streets of Europe. Imagine what the population of Britain might have been in the census of 1901 had not tens of millions, predominantly young and fertile citizens, emigrated to the colonies. Even despite that (and despite the famine in Ireland) the population of Britain had risen from 11 million in 1801 to 38 million in 1901, and then on to 59 million in 2001 and 65 million in 2015.

Startlingly, within the 65 million current UK citizenry reside (some always, many occasionally and a few in name only) more billionaires than in any other country in the world, some 117, who, combined, hold over £300 billion of wealth. Most, 72 of them, live in London. Britain has also deeper levels of poverty (and inequality) than anywhere else in western Europe. This is no coincidence: when greed gets out of control others are impoverished. That is the dynamic of uncontrolled neo-liberal economics. It has its cheer-leader: in 2013, Boris Johnson, mayor of London, stated that greed was 'a valuable spur to economic activity' and inequality was essential in order to nurture the 'spirit of envy' – apparently.

Uninhibited greed remains one of the greatest challenges to creating a more equal world. The realities of this have been clearly set out by French economist Thomas Piketty in his 2013 work *Capital in the Twenty-First Century*. What Piketty demonstrates is that wealth has a long-term tendency to accumulate upwards – into fewer and wealthier hands. The reason for this is that the rate of return on capital employed in rented assets (property, finance and equity) is, on average, greater than the return on productive work. Or, to put it another way, the already-rich get richer, and this dynamic, if not halted, is passed from generation to generation. As we previously stated, what matters most is where you were born and to whom.

Piketty is not a radical – although some have offered up that epithet to him – and he is not arguing for the tearing down of capitalism. His key prescription to halt the remorseless process of the concentration of wealth in the hands of the few is to implement a wealth tax at a global level. He sees this as an alternative to war or as a way of avoiding a series of systematic collapses of the global financial system. His thinking has been derided as being impracticable and utopian, but his suggestions are far more rational and possible than perhaps we might initially think, especially given that the alternatives to such thinking are increasingly dystopian in outlook and consequence.

Certainly there is a great distance to travel to a global society where there is intervention in the market built upon the rule of law – universal, transparent and fair. Yet this is the direction we are heading towards, mostly in small incremental steps and occasionally through significant surges forward.

It cannot be denied that the growth of trade facilitated by globalisation – the opening up of markets, the greater availability of capital, better transport and telecommunications, the exploding size of the Internet and the near ubiquitous availability of wireless technology (from the Americas to increasingly across Africa) – has transformed our world.

Furthermore, it has lifted hundreds of millions, perhaps billions, out of absolute poverty, even though the threshold for absolute poverty is set too low. Between 2005 and 2011, 300 million Indians saw their income rise above the $1.25 a day benchmark set by the World Bank, although mainly by only a very small amount. Over that same time period in sub-Saharan Africa the numbers in extreme poverty fell by over 100 million, while total population in that area of Africa was still expanding rapidly.

In the decade to 2015, at the same time that absolute poverty was being shrunk in India, relative poverty has been increasing, as is shown by a rising Gini coefficient (the standard measure of how unequal a society is) in both India and China. As the World Bank put it, all boats are not rising at the same rate on a rising tide. Across almost all the world the extremely rich are getting richer faster than any other part of society. In other words, the gap between the rich and the rest is growing, leaving increasing numbers subsisting in relative poverty.

Economically, mainly because a few are taking so much and more so recently, we are becoming more unequal. The super yachts are rising even higher above the horizon while the most unseaworthy of vessels carry thousands out of desperation across the Mediterranean, many sinking as they do so. If relative poverty in the world really is falling why do so many more people now risk their lives to try to escape from the poorest of countries? After all, migrants have a significant tendency to be among the wealthiest and most skilled to leave their country; they are the people who can raise the thousands of dollars to pay the predatory mafias who prey on their desire to make a better life for themselves in Europe, in North America, and in a few other wealthy enclaves of the world.

Proof of this growing global inequality – a reality predicted by Piketty's theorising – was the Credit Suisse Global Wealth Report 2014, which set out the estimated parameters of global economic inequality – 8.1 per cent of the global adult population held 82.4 per cent of global wealth. The global poor (people with wealth of less than $10,000) made up 69.3 per cent of the global adult population. Altogether they had amassed only 3.3 per cent of all global wealth.

The campaigning charity Oxfam spelled it out even more starkly after examining the same data. By 2014 the wealthiest 85 people in the world had assets worth $1 trillion – the same as the total wealth of the poorest half of the global population, some 3.6 billion people. This, Piketty argues, is a trend without a happy end if there is not the political will first to put the brakes on and then redistribute that wealth towards a greater economic equity, both within nations and between nations.

The researchers employed to produce the Credit Suisse Global Wealth Data Book have only limited access to the secretive world of wealth. What even these Swiss bankers know about who holds what wealth, and where, will be constrained due to the increasing complexity of global ownership structures, tax jurisdictions and the burgeoning legitimacy of aggressive tax noncompliance – evasion and avoidance. On occasion we glimpse behind the veil of secrecy of modern wealth – for example, in 2014 the *Guardian* newspaper revealed the scale of Chinese use of offshore banks, and in 2015 it was revealed that accounts in HSBC's Swiss bank were being used for significant amounts of illegal tax fraud across many European nations and beyond.

Even after seven years of austerity in the United Kingdom and falling real incomes for the majority of the nation's population, the UK is still placed 14th in the United Nations Human Development Index ranking, but it looks globally far more impressive for the UK and its citizens if we go for the economic jugular, the mean average wealth of an adult citizen in the UK, which is estimated to be £190,450, joint eighth in the world with Denmark. However, this is the mean average – that is, what it would be if wealth were shared out equally. Take the median average, the middle point in a distribution from lowest to highest, and that figure becomes £85,150. In Belgium mean average wealth is less than 2 per cent higher than in the UK, but the median average is over 30 per cent higher. Why the difference? It is the result of the scale of inequality in the UK. Inequality may be a little greater in the US than the UK, but it is far lower in Belgium than in the UK.

Some things matter more than money. Across the world water – its availability and quality – is increasingly the resource that defines the capability of people to develop all other resources. Water is an unequally distributed resource, and the clearest way to appreciate this is to look at the cartogram showing precipitation and population produced by cartographer Benjamin D. Hennig and shown on pages 90–91. Here it is clear where the 'pinch points' of water supply are: Cairo, the Gaza Strip, California, Greater Lima in Peru and nearly all of Pakistan and Yemen. Such areas are not necessarily reliant on precipitation alone for their water supply (most of which is utilised by agriculture) but may also draw upon precipitation from elsewhere, transferred by rivers such as the Nile or Indus, or from precipitation from an earlier age stored in underground aquifers known as fossil groundwater stores.

We know that water is a weapon and it is used to divide people and to entrench inequality. In the Palestinian Territories per capita water use is a quarter of its use in neighbouring Israel, even though the two populations share aquifers and climate systems. In Pakistan author Moshin Hamid places water centre-stage in his state-of-the-nation novel *How To Get Filthy Rich in Rising Asia*. Part of the answer is to control water resources. In California in 2015 the worst drought for a generation emptied reservoirs and left farmers struggling to survive as they bored ever deeper into depleting aquifers.

The uncertainties that frame the impact of climate change are often used as a reason to do nothing – or very little – to mitigate and adapt to it. However, even here progress is to be seen. The principle of 'polluter pays' is gradually growing in ascendancy. A Green Climate Fund exists to provide capital – up to $100 billion by 2020 – for developing nations to engage in mitigation and adaptation to climate change. In countries with a water supply struggle, like Pakistan, that is going to be vital.

The UK government has its own International Climate Fund, which is broadly re-targeting funds from the International Development pot. It showed that some progress was possible even under a centre-right government. Both China and the US have signed commitments to reduce carbon emissions by 2030, and even some Republican senators have accepted that anthropogenic climate change exists and are now starting to think of free-market approaches to addressing the challenge. In June 2015 the G7 group of nations, led by Germany's Angela Merkel, a centre-right politician, pledged to phase out fossil fuels globally by 2100. Many countries, some within the G7, have more ambitious targets.

Our optimism should be tempered by the possibility that this may all be too little, too late. The worst-case climate scenarios of a 6°C or even 7°C average global temperature rise by the end of the century could endanger the very existence of humanity. The probability is that this will not happen and that a better, but still problematic, outcome is more likely. Fortunately, it turns out that the more we are able to reduce economic inequality, the better the environmental outcome is likely to be.

Growing global economic inequalities are fast creating a dysfunctional, neo-liberal capitalism, increasingly unconstrained by regulation and democracy. This is a capitalism that is widely out of touch with the environmental, social and, particularly, the energy realities that frame all human endeavours. As Fatih Birol, executive director of the International Energy Agency, said in 2015, 'we need a peaceful divorce' between economic growth and ever-rising carbon emissions. Everywhere new thinking is edging on to the big political agenda, from both the grassroots and from some of the most influential 'global managers', be they from business or politics.

Fortunately, we live in a world of rapidly increasing

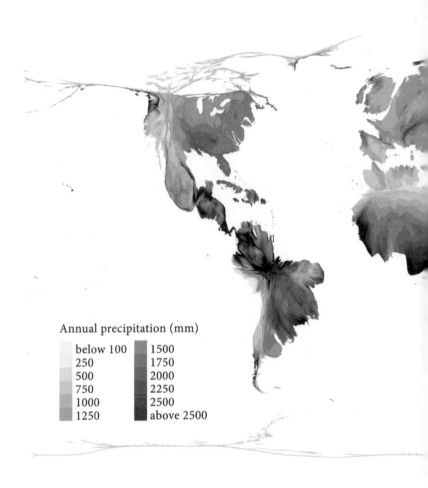

Annual precipitation (mm)

below 100	1500
250	1750
500	2000
750	2250
1000	2500
1250	above 2500

This map represents the world where most people live (area is proportional to population), and annual precipitation is overlaid on it. The lighter shaded areas are places that, on average, receive less than 250 mm of rain a year, whereas the darkest areas regularly experience precipitation over

3000 mm annually. By studying the relationship between these two variables, population and precipitation, it is possible to identify those locations where the water supply challenge is at its most potentially acute. The challenge appears particularly problematic around Cairo, Gaza and Pakistan.

equality in the access to knowledge and also access to the education required in order to process such knowledge. We are hopefully moving away from the scenario presented in the 2009 film *The Age of Stupid*, in which, in 2055, a single caretaker of a repository of knowledge watches archive footage of how the world had come to a dystopian present. Geography is a subject that provides the framework for inquisitive and imaginative humans to search for solutions to the challenges that face us. It is not a dystopian discipline.

Geography is essentially an action subject. It is one informed by an understanding of the environmental parameters of biodiversity, climate and morphology to enable a greater grasp of most current human dilemmas. Such dilemmas grow in complexity and inter-connectedness as we move towards a world of 10 billion people. There is a pressing need to move on from the fossil fuel world that has shaped our modernity and that fuels our economic inequalities, allowing the wealth of those who already have the most to rise to new, unprecedented and unsustainable global extremes. That time is coming to an end. It has to because it is not sustainable.

4

SUSTAINABILITY

It is a peerless English summer's Sunday morning – an uninterrupted blue sky and a gentle cooling breeze. Outside families and friends gather, sipping coffee and thinking about lunch. Washing hangs drying, and windows are thrown open to the radiant light. Then comes the billowing acrid smoke stretching out across a line of back gardens. A neighbour has decided to dispose of his old sofa by burning it in his back garden. Constructed out of mainly synthetic materials, the sofa burns well. Recently relaxing neighbours choke, clean washing is tainted and windows are hurriedly shut. The offending neighbour cannot work out what he has done wrong – it's his garden, his sofa, his choice – and we (his neighbours) are ganging up on him. An argument breaks out, and the police are called but are busy – resentment smoulders like the carcass of the burned-out sofa. Sunday is ruined, but the sun continues to shine.

What are we at liberty to do? Are we free to burn polluting old sofas stuffed full of toxic substances? Are we free to pass on the polluting externalities, the unpriced and uncharged consequences of such spontaneous house clearance upon our neighbours? Do we have a right, fundamental or otherwise, to soak up the sun unimpeded by the actions of others? Turn that radio off, stop the buzz of the mower and quieten down that child. Where do we start? Where do we stop?

Such are the fundamental dilemmas that all of us face in respect to our relationship to the planet, its biodiversity and our fellow humans. What are we at liberty to do, to each other and to our planet? It used to be much easier. Once upon a time most of us believed some very simple stories, even though we had to be told these stories because most of us could not read or write.

In the Book of Genesis, when God creates the first man and woman, the deity is explicit when instructing Adam and Eve of their place in the scheme of things: 'Be fruitful and multiply; fill the earth and subdue it; have dominion over the fish of the sea, over the birds of the air, and over every living thing that moves on the earth.' Dominion over the natural world was humanity's God-given right – the subduing of every living thing to the will of humans.

No doubt more environmentally enlightened readings of the Bible now exist, as they do for the Quran and the sacred texts of all the major religions. However, an assertive, dominating approach to the natural environment has been the hallmark of the move to agrarian and later feudal medieval societies. This did not change with the development of rapidly urbanising, capitalist societies, which were driven by the industrial revolution, colonisation and the exploitation of fossil fuel energy.

The last time population growth was as rapid as during the industrial revolution was during the neolithic revolution, when humans first settled into villages. At that time the oldest established religions and their rules eventually helped to make that transition from hunter-gathering more sustainable by helping to set parameters by which more densely settled humans could interact with each other sustainably beyond a social contract based only on our individual self-interest. But religions take time to develop. The one new religion of the current industrial revolution might be thought of as science, the modern 'way, truth and light', and as concerned with sustainability as any older region was and is.

An example of sustainability is the way in which in Hinduism the cow became taboo – some might say sacred. The cow was an incredibly useful addition to farming, providing dung for fertiliser and fuel, processed butter for lamps, milk for nutrition and power for ploughing fields. As such, its protection stabilised communities that were otherwise often persuaded to slaughter the animals in times of hunger. Even today, many rural families in India have at least a single cow that is often afforded a status similar to that of family member.

We are now in a new transition, from agricultural subsistence to industrial plenty, moving from village to city. This may seem overstated until you consider that in 1950 over 70 per cent of the global population lived in villages, and today that village population share is at 47 per cent and rapidly falling. This is partly because in recent decades overall global population numbers have expanded at a rate last seen only during that original many millennia old agrarian transition to farming. It is also because growing numbers of people are giving up village life around the world.

In the last two centuries humanity's impact on the planet

has been of such magnitude that almost everything around us is defined by our own activity. The Anthropocene – our age, the age when humans dominate – is just one example of such narcissism but also of a new reality. Such domination has had significant impacts on biodiversity. However, if humanity does crowd more into cities and away from the countryside, biodiversity may eventually benefit. But a great deal has to change first.

Biodiversity is currently declining, leading to a sixth major extinction event – that is, an event on a scale comparable to the last major extinction 65 million years ago that wiped out the dinosaurs. This occurs as humans plunder an increasing proportion of the global biomass, often indirectly, and squeeze what diversity is left into smaller and smaller islands of wild habitat. Chemical alterations to our atmosphere as a consequence of our insatiable urge to burn fossil fuels are having an effect on every square kilometre of our planet, from the highest peaks to the lowest ocean floors.

The scale of air pollution afflicting Chinese cities is now so great that barely an urban citizen remains unaffected in their daily life. The soil on which we invest our future washes off degraded lands across the world in unprecedented amounts and often in areas that have until very recently been humanity's most productive breadbaskets.

For very many generations the Midwest of the USA and China's Yellow River valley sustained large populations, but now soil erosion is a great threat to both areas. Even if we managed to get a grip on our current destructive tendencies, we clearly have much more to learn, and there will be harm that we are doing in other ways to the environment around us that we often barely recognise as damaging today.

The speed of current change should not be underestimated. As this chapter is concerned with sustainability, we must consider again our use of fossil fuels in this part of the book, even though we have touched on the subject earlier. In ancient times people had scant knowledge of fossil fuels such as oil. The Chinese drilled for oil from the 4th century CE, the inhabitants of the ancient city of Babylon used pitch from the Gulf oil fields, and the Persian physician and alchemist Muhammad ibn Zakariya al-Razi first distilled petroleum in the 10th century CE. However, the initial utility of oil was limited because it was difficult to transport and was not linked to the idea of motive power until the 20th century.

Less than 600,000 tonnes of oil were consumed worldwide in 1860. By 2010, consumption had reached 4 billion tonnes a year. And that's just oil – we could consider coal and, more recently, natural gas, and we haven't even mentioned shale oils and tar sands, possibly the last low-hanging fossil fuel fruit we could plunder, especially by means of 'fracking'.

The massive increase in energy use by humans has been of immense practical use, but we now know that it has also come at immense long-term costs. Converting stored energy into thermal energy has significantly changed the chemical composition of our atmosphere. Is such behaviour in any way sustainable? Can it go on? What are the moral imperatives that concern us when we are doing something today, the consequence of which will be potentially harmful

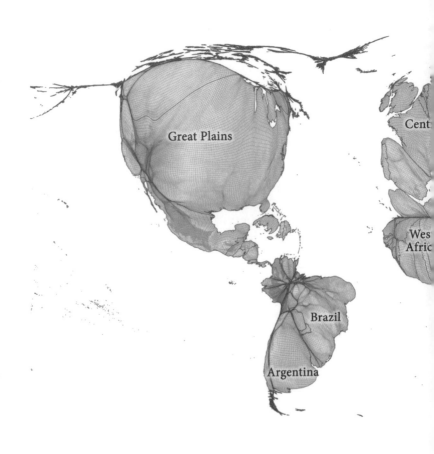

It is in the world's major croplands – the US Midwest, the Eurasian plains, the Deccan plateau in India and the Northern Plains of China – that much of the world's nutrition is produced. This is where the wheat, maize and rice that are today's main global food sources for humans around the world are predominantly grown.

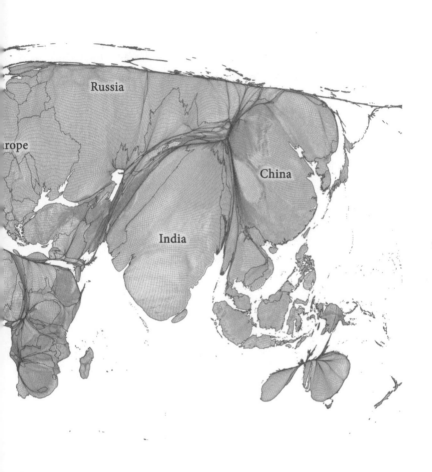

With two to three more billion mouths to feed in the next few decades, maintaining the ecological stability of the soil and climate in these biologically productive areas is essential. This map shows the planet drawn with the area of each small plot of land sized in proportion to the amount of food crops currently produced there.

to somebody tomorrow or maybe many decades in the future? These are the fundamental issues behind the idea of sustainability.

Sustainability has become a core issue for geographers in schools and universities only relatively recently. Of course, the idea that any environmental system could be both productive and remain diverse and healthy has been at the forefront of land and resource management for centuries. Woodland has been actively managed as a sustainable resource in many parts of the world through the necessity that such woods provided not only the majority of some local population's energy resources but also construction materials and even food resources. However, it was the thinking that grew out of environmental concern through the 1960s and 1970s that culminated in the 1987 UN-commissioned report 'Our Common Future' (the Brundtland Report). The report set out the expediency for the idea of sustainable development and signified a change in global human opinion.

At the heart of sustainability is an idea of inter-connectivity between different domains: ecology, economy, politics and culture. Geography is central to pinning these domains together, so that a bigger picture can be envisaged – from fluvial hydraulics (water supply, droughts and floods) to demography (population size, composition and distribution).

At the sharp end of thinking about sustainability is political philosophy – that area of thought that examines power, authority, legitimacy and justice. Without a coherent political philosophy, sustainability could be a rather marginal intellectual idea. However, with a political philosophy sustainability becomes central to humanity's progress and development – or survival, if you take a more pessimistic perspective.

The 20th-century US political philosopher John Rawls constructed a thought-experiment to demonstrate that universal social justice is the rational path for humanity to pursue, both within and between individuals, societies and generations. Rawls implies that sustainability is the long-term environmental perspective of social justice. In his thought-experiment a perspective entitled 'the original position' is assumed. Participants are asked to imagine they are as yet unborn and know nothing of the life they are to be born into. This means they will hold no prior assumptions about gender, ethnicity, social and economic status nor any of the underlying assumptions about what might be considered a 'good life' in any society or social stratum. Any individual may be born disabled or not, poor or wealthy, male or female, in Africa or in America and so on. The number of such outcomes would reflect the probability of any outcome at present. So, if you are playing with this experiment in your imagination, you must remember that you have a far greater chance of being poor than rich. But you also have more chance of being able than disabled, at least when you are young.

Rawls's thought-experiment asks you to ponder what sort of society you would like to be born into? Would you prefer a society based on uncaring competition that results in increasing levels of inequality – where those born with advantage retain this advantage simply by virtue of birth not merit? Or would you prefer to have a society where equality of opportunity and outcome is sought after and the merit of each individual is valued?

Rawls argues that the rational choice is a society where equality of opportunity is open to all, that the greatest benefit should be afforded to the least advantaged and that liberty should be equal and equally available to all. To choose otherwise, Rawls proposes, would be irrational. Others go a little further than Rawls and suggest that in many areas of life 'equality of outcome' is just as important – for instance, in the outcome of successfully surviving your first year of life as a child.

If we extend this experiment a little and invite you to project into the future and consider what sort of planet you would like your great-grandchildren to be born into and in which they should try to live well if they were to have no prior knowledge of the place and position into which they were to be born, what would you wish for them? What distribution of possible chances would you desire? As world population stabilises the average person will have eight great-grandchildren. Some of these eight will be 'winners' in life and others will be life's 'losers' – which possible outcomes would concern you the most?

Would a good future planet for your great-grandchildren be one denuded of biodiversity? A planet where the basic environmental resources such as soil and fresh water were more limited? A planet where wild fluctuations in weather patterns resulted in a wholly unpredictable climate? This would be a planet where the increase in atmospheric and oceanic temperatures had made the lives of billions – if not all of us – more challenging. This would be a world of intense inequality and the denial of basic liberties to the majority. However, if you had no children or grandchildren would you care?

At the other end of the scale of possibility would be a world where your great-grandchildren would enjoy a planet that contained just as much wonder and beauty as it does now or even more. Biodiversity would be flourishing. Equality of opportunity would be afforded to all, regardless of where and who your descendants found themselves to be. The challenge of climate change would have been addressed by the more equitable utilisation of energy resources, material resources and capital. Life would not be appreciably worse than that which many enjoy today, and in many ways it could be better: we would be moving towards a sustainable lifestyle. And if you, yourself, did not have offspring or your children did not, you might still care for a wider humanity, as if others were your relatives; not least because they are.

Here is the conundrum of sustainability: to achieve a sustainable world we have to try and factor in the future, to try and pick choices that will not excessively make the options available to our grandchildren, and everyone else, more constrained overall and so to enable them to have a planet that is able to provide them with the ability to live fair, equitable and fulfilling lives. You might think it is not much to ask, but you would be wrong. Stacked against such aspiration are most vested interests in the political, cultural and economic world that we currently inhabit: an economic system based on the aspiration of continuing and unending economic growth: the 'culture of more'; the 'politics of greed'; a world where the political philosophy of John Rawls is yet to inform our mainstream political view, let alone even more progressive viewpoints.

Try this second simple thought-experiment. Given the choice at the ballot box would you vote for policies that probably made you materially poorer, intervened in your ability to express unhindered choice and, although you would not experience much direct benefit from them, would be of great benefit to your, as yet unborn, great-grandchildren? Or would you vote for policies that maintained your current standard of living for a few more years and from which you could see immediate benefit, that allowed you to have more and to have more choice in having more, but would be of significant and potentially catastrophic harm to your, as yet unborn, great-grandchildren?

At this point you may be thinking that geographers have a political agenda. You would be right, and that is what often dictates the areas they focus on. It has always been so. In the past, with a colonial perspective, geographers often focused on the exploitable possibilities of other lands. Today the most pressing challenges concern sustainability, globalisation and equality – all of which are inter-connected and framed by the environmental constraints imposed on us by the energy flows through our biosphere.

The key problem of introducing the idea of sustainability into the political process is how to make sustainability fit with democracy. Sustainability is currently progressing towards the forefront of political thinking in many countries, which have a variety of political systems, from centralised, post-socialist China to the neo-liberal citadels of the USA and UK. Everywhere there is still resistance to the very idea of sustainability from many political and business interests, although often the greatest challenge is disinterest, apathy and resistance to individual behavioural change. Yet a huge number of politicians representing almost all nations are today unambiguous: the future is about sustainable resource use. Indeed, targets have been set. We have now reached the stage of the haggling over how we will get there and how soon, not over where we should be aiming to get to.

How do we begin seriously to consider how to manage the relationship between the human aspiration for more and the planet's inability to fulfil such a desire, at least not for most of humanity? Mike Berners-Lee demonstrates just how big a question this is in his mind-bogglingly instructive book *How Bad are Bananas: The Carbon Footprint of Everything*. What matters most is the relative impact of so much of our behaviour that we take as given, normal and everyday.

An average Australian citizen emits 30 tons of CO_2 a year. In contrast, an average Chinese citizen emits 3.3 tons. Furthermore, much of that Chinese contribution is then exported abroad and consumed in other places, including Australia. About the same amount of CO_2 is generated by flying between London and La Paz in Bolivia and back as a Chinese citizen generates in a single year.

A connected world is an energy-intensive world. Although an increasing range of alternatives exist to generate electricity, or to create heat and even to power cars, as yet no viable replacement for high-octane aviation fuel exists. If you were willing to expel 3.5 tons of CO_2 into the air, plus a bit more for the 730-kilometre journey south from La Paz, you would find yourself at the Salar de Uyuni. These immense salt flats in southern Bolivia are so large, bright white and cloud free that passing satellites use them to calibrate their visioning equipment. Alternatively, you could travel driving an electric car and keep your carbon footprint down. Such an electric car would be powered by a rechargeable battery within which would be up to 40 kg of lithium. Lithium is the lightest metal in the periodic table, silvery white in appearance and reactive immediately when it comes into contact with water or air. It is an essential component of all rechargeable batteries, and it is, therefore, an important resource for a low-carbon future.

Between 2000 and 2012, the world price of lithium carbonate increased threefold. The salt flats of Salar de Uyuni contain between 35 per cent and 50 per cent of the known world reserves of lithium. They are 200 kilometres and three hours' driving time from the silver mountain of Potosí. Geological good fortune often comes in geographical clusters, as in the case of Britain with its relatively easily accessible coal, gas and oil.

As yet, virtually none of Bolivia's lithium has found its way on to the world market. Its southern neighbours, Chile and Argentina, dominate supply. The problem is one of globalisation. All of the world's lithium mining companies are, bar one, Western owned: four Canadian, three Australian and one each from Austria and the USA. The largest lithium producer in the world, Talison Lithium, was Australian owned a few years ago; by January 2015, the company was 51 per cent owned by the Sichuan Tianqui Lithium Corporation of China working in partnership with the Albemarle Corporation of Baton Rouge, Louisiana, USA.

Bolivia wants to develop its lithium resource itself, especially given that the history of Spanish silver mining hangs heavy over the nation. The Bolivian government had been embroiled in long-running disputes with foreign companies that controlled the nation's core utilities, water, sanitation, and electricity. Recently, the government took these utilities back under state control and nationalised them. One British company, Rurelec, which owned just over half of the country's largest electricity company, Empresa Eléctrica Guaracachi, claimed $100 million in compensation but were negotiated down to accept $31 million. As we write, outstanding claims against the Bolivian government remain above $1 billion and are being argued through the International Centre for Settlement of Investment Disputes (ICSID), even though Bolivia left this institution in 2007. Apparently, the $800 million required to kick-start Bolivia's progress towards what the London-based *Financial Times* calls 'the Saudi Arabia of lithium' can be raised only through foreign investment.

Bolivia's government is relatively radical and 'leftist'; it is committed to economic equity for its people and environmental sustainability. Bolivia bears the scars of colonial oppression, including the scars on the physical landscape, to remind it of what exploitative resource extraction can lead to. On the other hand, it has been estimated that the Salar de Uyuni could provide enough lithium for 4.8 billion electric cars, although, given that the world population is slowing to a peak at around 10 billion, we should never need as many as 4.8 million cars in the world, even if they are all electric. Bolivia is a part of the world that is possibly key to a more sustainable global future, both environmentally and economically.

Today some of the most eminent scientists in the world suggest that environmental, economic and social existential crises are arriving with increasing rapidity. Of course, a few have always done this. The Rev. Thomas Malthus, the world's first economist to be salaried (by the East India Company), was chief among the early doomsayers. However, just because a few have always cried doom and, until now, have always been wrong, this does not mean that we should ignore similar calls. In fact, it might be because enough people listened in the past that doom has so often been averted and new insights gained. Malthus was completely wrong in his warnings about population growth, partly because there was so much he did not know as he put down his ideas at the end of the organic energy age, back in 1798.

Today Lord Martin Rees, Emeritus Professor of Cosmology and Astrophysics at the University of Cambridge, past President of the Royal Society (as establishment as you can get) preaches doom. Reminiscent of Marvin, the paranoid android in *The Hitchhiker's Guide to the Galaxy*, Professor Rees suggested in his 2003 book *Our Final Century: Will the Human Race Survive the Twenty-first Century* that humanity's chance of successfully negotiating its way to the end of the 21st century is about 50:50. He included challenges such as bio-terrorism and out-of-control nano-technology in his smorgasbord of impending doom.

Rees is far from alone. Even in China 'green cats' within the Communist Party and beyond, such as the writer Ma Jun, now express serious concern about environmental pollution choking China's dash for growth. Concerns about pollution are widespread within the Chinese public consciousness. More than 150 million people downloaded investigative reporter Chai Jing's documentary *Under the Dome* about the acute threat air pollution posed to human health and the environment. This number of viewings occurred in less than a week, but at that point the Chinese authorities took it off the Internet in March 2015. In that same week it garnered 500,000 YouTube views outside China.

Across the geographical tradition researchers all over the world are collecting data that will reveal the impact that humanity is having upon the natural world. Their measuring includes the bubbling methane being released from a warming Arctic Ocean – a global warming tipping point that some scientists have suggested could be crucial. Some are attempting to calculate the loss of soil as a consequence of different farming practices, something of supreme importance considering the increase in food productivity that will be required in coming decades. Geographers worldwide are looking into cities, migration, air pollution, biodiversity loss, hyper-mobile capital, transport and so much more – it would be easier to ask what isn't under the research lens of geography. And all the time these thousands of researchers are trying to join together the disparate information they, and so many others, gather.

The interconnectedness of geographical processes across the human and physical spectrum can be illustrated by innumerable instances. Two examples follow, one from over two centuries ago and one that is far more contemporary and concerns thinking about the future.

In 1783 the Laki fissure in Iceland erupted, pouring lava across the land and pumping poisonous clouds of hydrofluoric acid and sulphur dioxide into the northern atmosphere. A quarter of Iceland's population died, and the cataclysm spread way beyond that isolated island. In Europe thick smog blocked out the sun – in the height of summer – and food crops failed. In Witney, Oxfordshire (incidentally the Parliamentary seat of the UK prime minister at the time of writing), the visiting Methodist preacher John Wesley observed that the unremitting thunder storms with lightning puncturing the dense fog led many to think 'the day of judgement had come'. He noted that he had never seen churches so full.

The UK weather in 1783 didn't clear easily and the high-altitude sulphur dioxide affected food production in Europe for another five years. Winters were bitter, followed by major flooding in river valleys and then a mixture of summer droughts and intense hailstorms, all of which decimated crops. In 1789 the French revolution was partly set in train by desperate food shortages as a consequence of a decline in agricultural productivity.

The 1783 event could be seen as a relatively local example of what might occur as the climate becomes increasingly erratic throughout the 21st century. The Laki fissure in Iceland remains active, with a neighbouring volcano,

Eyjafjallajokull, providing both a challenge to newsreaders across the world and disruption to air travel right across Europe in 2010. These eruptions demonstrate the inter-connectedness of physical geography with human, economic and social processes.

The second example is one that is yet to have a profound impact upon humanity, but it is on such a scale that it is probably not even fully comprehended to date – it is the Himalayan glacial retreat as a consequence of global warming. This retreat will affect the water resources available in the most densely populated areas of China and neighbouring countries, and the mismatch between available water resources and development aspirations is stark. Chinese geographers have recently reported on their measurements of retreating glaciers. Professor Yao Tandong is Director of the Institute of Tibetan Plateau Research, and he estimates that by 2060 two-thirds of all Himalayan glaciers will have disappeared. The consequences of this on both the Yangtze and Yellow River basins are profound, and contemporary fieldwork research by Wu Shanshan on the upper catchment of the Yangtze already confirms changes in flow patterns.

Whatever happens upstream in a drainage basin affects the downstream processes, whether it is increased sedimentation due to changes in rainfall run-off as land use patterns alter or variations in the overall flow due to changes in climate patterns. The Chinese authorities are more than aware of this dilemma and have thrown a staggering quantity of resources at attempting to address the long-term impact that dwindling water supplies will have on China's productivity and economic well-being.

Water has become an issue of fundamental importance to the Chinese – posters adorn city walls in China stating that 'the last drop of water on Earth will be a human tear' and exhorting the populace to conserve water for the common good. To a large degree the key problem is entirely geographical. There is simply a mis-match between where sufficient and even excess water is found – the Yangtze River basin – and where water is scarcer – the colder and more arid northern regions of China, drained to a large degree by the Yellow River. Mao Zedong first recognised this dilemma back in 1952, and six decades on the first spurs in a giant civil engineering project to transfer water from southern China to northern China are operational. A pumping station in Beijing came online in December 2014, providing 500,000 m^3 of water daily for up to 5 million citizens.

The scale of China's current geo-engineering endeavours is immense. To date $81 billion has been spent building two canals, one along the eastern seaboard from Yangzhou to Tianjin and another, more central, canal from Danjiangkou Reservoir to Beijing. The latter of these two canal routes covers a distance of 1,273 kilometres and descends some 99 metres, an average gradient of 1:25000, with gravity providing the energy to move the water northwards. Over that journey the canal crosses 205 rivers, 42 railway lines and is crossed by 735 highway bridges. Some 300,000 people have been displaced, primarily as a result of the construction of the Danjiangkou Reservoir. Yet, even with all this application of science and technology, it is doubtful whether the south–north water transfer scheme is a sustainable solution. Even the Beijing head of the project, Sun Guosheng. stated:

'To save is to survive. Individuals and businesses should all use the tap sparingly, and heavy fines should be levied for wasting water.' To this end the Chinese state has introduced differentiated water charges to try to use market forces to reduce consumption by making individuals, agriculture and business more price-conscious in their use of water.

While it is true that the earliest human settlement in the Tigris River valley increased its energy efficiency by employing irrigation systems, it is also the case that humanity has never tried such large-scale hydrological management schemes as are currently being attempted in China – and probably never will again. China is approaching its population peak, currently estimated to be 1.375 billion people in 2023. After that it is set for rapid population decline – that, after all, is what having had a one-child policy for decades eventually results in.

The Chinese case does, however, illustrate the scale of the sustainability challenges that face the world. It also demonstrates that national-scale solutions can rarely, if ever, be achieved through the use of the market. The capital required, the huge lag time in returns to capital employed and the necessity of having a fast-track planning system that is able to over-ride individual and community rights are factors that work against the private sector's ability successfully to carry out global geo-engineering projects.

For geographers, problems such as water, pollution and population bring together all the elements of the subject: historical and anticipated changes, planetary change, local impacts, geo-morphological process, tectonic adaptation, hydrological and climate dynamics, all wrapped up in a socio-economic context within a political framework.

What to do when you've never done anything as big before is not a uniquely Chinese problem. Take large-scale engineering and infrastructure projects such as the UK's high-speed rail development, Nicaragua's trans-oceanic canal or the 57-kilometre Gotthard tunnel, which will take high-speed trains completely under the Alps when it is completed in 2016. These are all engineering projects of an immense scale that involve a range of geographical complexities, from geology to economics. All such projects require the requisition of huge energy inputs, and they often encourage greater utilisation of energy than was the case before, albeit far more efficiently for the work being done and the result produced.

There have been immense achievements over time through the application of human ingenuity. Humans have reshaped land, built immense walls, demolished mountains, marshalled whole river basins to their will and even hauled giant stones long distances to construct monuments to their wonder with nothing more than sheer muscle power. Clearly, investing time, capital, energy and ingenuity into collective projects, where the benefits were widely shared, has a long and geographically dispersed history.

One of the most successful – and at times problematic – human adaptations of the environment has been the re-positioning of plants. This process has been carried out from the very early days of human settlement. Everything originated somewhere and at some moment in time, but increasingly both plant and animal species are dispersed beyond their original environments. This dispersal has been of huge benefit to humanity, with grains from greatly adapted grasses providing most human and domesticated animals' global nutrition, and generations of hybridisation and selection producing many new varieties of plants and animals.

Introduced (non-native) crops and livestock account for 98 per cent of all food produced in the USA, which is the world's largest exporter of food. Kansas is the largest wheat producer in the USA, and scientists funded by agricultural interests and working at Kansas State University predict that for every 1°C rise in average temperatures a 6 per cent decline in wheat yields will occur in their state. Climate change may be the most significant spur to moving biodiversity around the planet ever seen and with potentially staggering – in the geological and evolutionary sense – rapidity.

In recent years an estimated $27 billion has had to be spent annually in the USA on controlling invasive species. That cost should be set against the $144 billion value of US agricultural exports in 2013 and the far greater domestic worth of those crops, which in human terms are of almost infinite value.

Of course, ingenious and inventive endeavour, as set in train by the colonial plant hunters, is not without cost and miscalculation. Indeed, the range of examples of non-indigenous species wreaking havoc on eco-systems is extensive. Biologists have always found it most instructive to observe such changes on islands that were, until relatively recently, remote and disconnected. In places like Hawaii introductions of new plants have, contrary to popular belief, led to greater plant diversity than existed naturally. In Hawaii, 12 additional non-native plants have been added for every native plant lost, although other biological species, such as snails, have experienced a decline in their diversification – fewer snails are introduced from other places to replace snail species lost.

Although great costs have been incurred in controlling alien species – such as rhododendrons in Wales's Snowdonia National Park – the same can also be said about controlling native species, including bracken in that same Snowdonia National Park, for example. The idea of banning the intro-duction of alien species under any circumstances because of possible negative impacts on eco-systems is rather like trying to stuff the genie back in the bottle. We are way past that point.

The UK imports over £1 billion of ornamental plants every year: the flora of the Fynbos in South Africa, the Mediterranean, Australasia, China and the Americas populates the gardens of Britain. London parks play host to ring-necked parakeets from the Indian subcontinent. There is barely an urban space in Britain that doesn't have a buddleia (otherwise known as the butterfly bush) wafting its enticing fragrance across the summer, a far superior introduction from China in the 19th century than the opium that was proffered to the Chinese by the British in the other direction.

Geographically speaking, it is not sensible to be simply dismissive of the impact of alienisation. Clearly there are impacts, costs and consequences, but these have to be set against the positive benefits and the general increase in biodiversity and eco-system resilience that new species can also provide. Geographers probably see the bigger picture better than more fixed-focus eco-system biologists do, although British plant biologist Dr Ken Thompson provides a notable exception to that generalisation in his entertaining book *Where Do Camels Belong?* (2014).

The greatest threat to biodiversity loss remains habitat loss from encroaching agricultural monocultures, spreading deserts or urbanisation. The use of extensive areas of natural landscape has accelerated as the population of the world has expanded, and the need to grow more and more food is paramount. Hence, the dilemma arises as to whether yields can be expanded from existing land or whether non-agricultural land – or even those wild habitats where the majority of existing global biodiversity resides – will have to be farmed.

Set against some optimism that we can feel about the future must be set the reality of desertification, often as a consequence of population pressure and vegetation removal. What the UN labels as 'dry lands' account for 34 per cent of the world's terrestrial surface. Such areas will probably expand with climate change. Also to be considered are the pesticides, the technological and energy inputs into agriculture that have been largely responsible in the late 20th century for expanding yields and our currently insufficient overall global food supply. Can such energy inputs be maintained let alone expanded? Will insects evolve that are resistant to all our poisons?

The agricultural 'green revolution' – where bio-technological, mechanical and organisational innovation came into play – has resulted in dramatically greater yields from cereal crops worldwide since the 1960s. This may have been essential to supporting economic and population growth globally, but further improvement – the extraction of more and more food energy from a fixed biological potential – is required in order for economic growth to continue apace into the future. Poor soil management, over-cultivation and changing climate conditions have all contributed to the increasing loss of productive topsoil across much of the world.

A sustainable planet may only result from humans moving into new technological terrains. There is still much debate to be had about controversial genetically modified (GM) organisms, which have sometimes been suggested as a panacea for our global nutritional dilemma. This debate rumbles on and will not be resolved in the near future. Public concern has resulted in the indiscriminate use of the precautionary principle, and GM crops are banned in many countries. No such concerns, however, have restricted fracking or the extraction of shale oils. Some energy resources have more political and economic sway than others. Things can also suddenly change.

In a globalised world no individual, community or society is an island, isolated from the realities of global environmental change. Sustainability is a global imperative. Unlike wobbly political rhetoric, this really is something where we are all in it together. The question remains, however: how can people be persuaded of that reality, especially those with the most power and wealth?

Little by little, action to mitigate and adapt to climate

change is under way. There are growing movements for the conservation of biodiversity; strenuous efforts are being made to expand the food supply and improve its resilience, as well as the beginnings of a truly momentous energy-generation revolution, in which we move away from the fossil fuel age into a low- or even no-carbon one. Everything is in play. There will have to be an acknowledgement that until some of the fundamental challenges of human economic inequality are addressed, progress will be obstructed.

Sustainability and the debate it creates are where the most radical and, indeed, revolutionary political responses are being framed. Geography could be and should be at the forefront of these exchanges. But how could that be done?

In the 1970s, one of the most prescient thinkers was the social philosopher André Gorz. Gorz grasped the impact of technology on productivity and the consequential impact on humanity, and he also foresaw the need to marry politics and ecology. He understood how energy capture defined the parameters of human endeavour.

Gorz also propounded the idea of a universal basic income. He was not the first, nor the last, to do this. Indeed, here lies an idea that has huge sustainable merit in its potential for providing citizens with the basic means to support their lives and thus allowing them to be partially liberated from the pursuit of more, or at least those citizens who would choose this low-impact and leisure-rich lifestyle.

To become more sustainable requires addressing gaps between us. Economic inequalities make growth appear attractive, but it is now becoming obvious to the majority who have even a slight numeric ability that economic growth cannot continue unbounded and without consequence on a finite planet. In our fully globalised world we can now see the consequences of not sharing better and exploiting less.

An approach such as a citizen's wage is an example of confronting the challenges and exploring new ideas and possibilities in a practical and realistic manner. This kind of imaginative response lies at the heart of the new geographies that will become an invaluable tool for humanity to map a route through the 21st century. Sustainability is about survival, but it is so much more than that: it is about creating a better, fairer, more convivial world where 'affluenza' (always wanting more) and the destructive accumulation of the world's environmental resources can both be significantly tempered. It is about the world you would like your great-great-grandchildren's generation to inherit. It is about hope, not fear.

5

MAPPING THE FUTURE

In a connected world that is saturated with information and news, from terrorist killings and air crashes to collapsing economies and drowning migrants, one significant piece of good news in June 2015 pierced the otherwise continuous digital diet of doom. At a Bavarian stately house nestled at the foot of the Alps, the G7 nations – a group of seven major advanced economies and the EU (which together represent almost half of global GDP) – gathered for a two-day conference. There was much to discuss: Russia's foray into the Ukraine had upset one part of the current precarious geo-political balance, Greece's economic problems highlighted the instability in rich-world economics and terrorism (claiming to be Islamic in conception) added to the bulging in-tray of issues. However, from this meeting came a statement that was, potentially, historically monumental. Germany's chancellor, Angela Merkel, stated: 'We aim to decarbonise the global economy in the course of this century.' The communiqué, which was signed by all seven leaders, affirmed: 'We are committed to achieving an ambitious, people centred, planet sensitive and universally applicable Post-2015 Agenda for Sustainable Development that integrates the three dimensions of sustainable development – environmental, economic and social – in a balanced manner.'

Then, within a fortnight of the G7 meeting, Pope Francis published an encyclical (papal communication) entitled *Laudato Si': On care for our common home*. This directly addressed climate change and radically challenged the broader, unsustainable dynamics of an economic system predicated on constant economic growth. The pope was not reserved in the language he used; he understood the inter-connectivity of the environmental crisis with the wider economic world. In an astonishingly radical critique of the economic philosophy that has brought the human world to where it is today, Pope Francis stated:

> Politics must not be subject to the economy, nor should the economy be subject to the dictates of an efficiency-driven paradigm of technocracy. Today, in view of the common good, there is urgent need for politics and economics to enter into a frank dialogue in the service of life, especially human life. Saving banks at any cost, making the public pay the price, forgoing a firm commitment to reviewing and reforming the entire system, only reaffirms the absolute power of a financial system, a power which has no future and will only give rise to new crises after a slow, costly and only apparent recovery.

In the summer of 2015, in the run-up to December's climate change negotiations in Paris, it became widely acknowledged by the global elite that the era of fossil fuels was over. Three years before any of the leaders of the richest parts of the world added their voices, Xi Jinping, when he became president of China and general secretary of the Chinese Communist Party in 2012, spoke plainly about how he saw the sustainable future for the Chinese nation. He stated: 'Our people love life and expect better education, more stable jobs, better income, more reliable social security, medical care of a higher standard, more comfortable living conditions, and a more beautiful environment.'

The rapid growth of the Chinese economy, fuelled by an increasing consumption of coal and (mainly imported) oil, has been unprecedented. China is now the largest emitter of carbon dioxide in the world. Its emissions per capita are still a third of those in the USA, and much of this output is generated through producing things consumed elsewhere. Because of this, it would be naive to suggest that the blame for being the largest emitter can be placed solely at China's door. But how does China fulfil the aspirations of its people articulated by Xi Jinping, while simultaneously acknowledging that the growing Chinese economy will have a detrimental effect on the environment? The dilemma is how to manage a transition to a low-carbon economy in a nation where, in recent decades, aspirations have risen even faster than living standards.

At the end of 2014 the USA and China agreed to set binding targets to address carbon emissions. The US promised to reduce emissions by a quarter by 2025, while China, acknowledging for the first time that its rising emissions were unsustainable, agreed to ensure its emissions were declining in absolute terms by 2030. To achieve this objective, China proposed to install 1,000 gigawatts of zero-carbon energy technology (wind, solar, nuclear and hydro) by 2030. These green-power plans exceeded America's total capacity for electricity generation in 2015. The scale of the Chinese commitment to renewable energy is just one further indication that we have now reached the

The world at night: the light from human settlement shines back out into space in this map on which land masses have been resized according to where people actually live. If any area is light on this map either the people living there have little or no electricity or they use what they have wisely enough not to burn it wastefully at night.

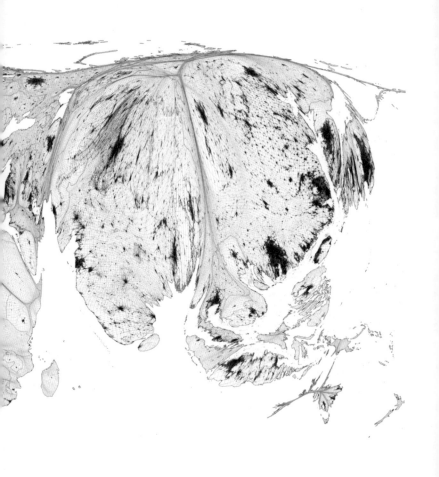

Tokyo, the Nile delta, northwest Europe and the 'endless city' that rings the Pearl River delta in southern China all burn bright. It is in such places that the transition to a post-fossil-fuel future will be most effective in reducing the carbon footprints of these efficient but over-consuming urban dwellers.

tipping point into the post-fossil-fuel age, a new epoch for humanity.

Climate change is the spur that has drawn the utopian gaze away from fossil fuel use and crude measures of supposed economic growth. Yet, enhanced global warming is but one of a wide range of inter-connected challenges that we now – collectively – face, both on a planetary and local scale. Some of these challenges will certainly become worse, including increasing desertification that destroys croplands, rising sea levels that inundate coastal settlements and escalating extreme weather events. No matter what action is taken today, too much damage has already been done. In other words, the fossil fuels have already been burned – their residue is already in the atmosphere – and the environmental consequences are still to unwind. This means that restricting global average temperature rises to a maximum of 2°C above pre-industrial levels – the benchmark 'obligation' agreed at the 2009 Copenhagen Climate Conference – might prove difficult, if not impossible, to achieve, no matter what we do now. Adaption then also becomes a key challenge, alongside rapidly reducing further pollution.

It is a commonly heard, if misguided, refrain that, of all the challenges we face, population growth is the most pressing. The TV naturalist and former BBC manager David Attenborough has warned that 'we are heading for disaster unless we do something', and he once even described humanity as 'a plague'. However, commentators like Attenborough are often very vague on what that 'something' would be. Was he thinking of something like China's one-child policy or the UK government's two-child welfare policy, by which tax credits and Universal Credit will be restricted to two children after April 2017? However, the poor in the world, including those in rich countries, pollute the least, so further restricting their numbers would be the least effective policy.

Statistics show that China's poor hardly contribute to pollution at all and that children born into the poorest families in the UK also consume the least. Preventing population growth among the poor is an inefficient mechanism for reducing carbon pollution. Furthermore, it must be recognised that the rate of human population growth has been rapidly decelerating since 1971, and the year in which more babies were born than at any time before or since – the 'peak baby' point – occurred in 1990. As we write, the global fertility decline is accelerating. People almost everywhere are having fewer and fewer babies.

The main driver of worldwide population growth today is ageing, not births. When life expectancy doubles, world population doubles. Yet, as a result of general health improvements, even with the rapid increases in life expectancy predicted worldwide, an absolute population decline is still forecast to occur within the lifetimes of those being born today. How steep this decline will be, and exactly when 'peak population' will be reached, remain questions that are actively debated among demographers, and the critical point will be easily brought forward or delayed through the slightest of changes in circumstances.

The real problem is not that humans are living a few years longer but the extent to which we each, varyingly, stamp our footprints across the world. What we each consume in total over our lifetimes is of fundamental importance. In other words, what is the size of our individual ecological footprint? Do we all begin to tread more lightly? Or is the reverberating stomp of a few still felt all around the world far more than the tread of the many? How does one super yacht and all it takes to build, equip, power and crew compare with the resource impact of a shack housing a family (or two families) in a shanty town?

Greater numbers of humans do not inevitably place more pressure on biodiversity. It is entirely possible to have more people on the planet than there are at present alongside a lower overall level of pollution. However, such a scenario cannot be realised with contemporary carbon-fuelled modes of production. Over the 20th century, biological productivity – the rate of generation of biomass – has doubled, partly because we have temporarily added fossil fuel-based fertilisers to so much of the soil. Over the same period the global human population rose approximately four-fold and our economic output increased 17-fold. Such a mismatch in growth rates was made possible only through the increasing use of fossil fuels.

Currently, humanity appropriates just over a quarter of all new biomass produced globally per year. To decrease fossil fuel use substantially over the coming century we must begin to better understand the biological constraints, especially if we and other species are to retain a large element of the wilderness in which biodiversity thrives the most. However, the need to increase global food production may place unbearable pressure on existing wilderness areas.

This necessity has been made all the more acute by increasing meat consumption worldwide.

Per capita consumption of meat has doubled worldwide since 1961, and producing meat for human consumption requires far more energy than deriving the equivalent quantity of food energy for vegetarians. Although a completely vegetarian world would be easier to feed, this is unlikely to be accepted. However, meat consumption is falling in nations such as Australia and Argentina, where it has historically been high. Many meat-eaters could accept eating meat just once a week, as many of their grandparents did.

What food is produced, how efficiently it is produced and whether we are able to use less fertiliser in future will be partly controlled by how well we manage available water resources. Declining snow-packs are resulting in waning river flows in many parts of the world. Furthermore, ground water is being over-extracted in many river catchments, from Saudi Arabia to California, placing additional pressure on water resources. Excessive usage of water is readily encouraged in the pursuit of 'aspirational' lifestyles, and the increasing demand for private swimming pools and pristine golf courses further depletes vital water sources.

We will not manage water successfully if we fail to make the transition from an individualistic to a collective mentality. Reaching the limit of sustainable collective consumption will force us to change as a species. Climate change is altering our belief that contemporary economics promotes equilibrium. Economists borrowed the idea of equilibrium from models of nature, but how nature is now being changed will affect economic thinking far more directly.

Even the world's immense oceans are being altered as a consequence of climate change, becoming increasingly acidic as they absorb more carbon dioxide from the atmosphere. This trend will bring substantial changes to marine eco-systems, and crucially it may significantly reduce fish stocks.

The challenges that face humanity are not confined to the bio-physical world. As Pope Francis described, it can now appear as if the world of global finance wields 'absolute power'. Underpinning the accumulation of capital (wealth) in fewer and fewer hands has been the development of digital financial trading in which the importance of transparency within a democratic framework of accountability has been eroded. In 2010, somewhere between $21 trillion and $32 trillion in funds was unreported, estimated to be largely untaxed private wealth sitting offshore.

It must be remembered that the enormous stores of monies currently being hidden by the super-rich around the planet are not literally 'offshore'. Digital connectivity has enabled wealth to be held in the UK in houses, investments and even government bonds but to be banked through tax havens, such as in sleepy Road Town, the diminutive capital city of the British Virgin Islands, and the many other 'treasure islands' of contemporary tax avoidance used by the extremely rich. This group are just the richest fraction of the 1 per cent who now own half the world's wealth.

The concentration of most economic power in so few hands has been further enabled by the drive to milk the maximum rent from capital, which has resulted in the sale of debt becoming more profitable than production. Work does not pay as well as renting assets, lending money for private consumption or buying government bonds. Often the capital employed to do this sits behind an opaque ownership structure and outside anything but the lightest of light touch regulation. Such are the ranges of grey between the formal and informal economies.

The iniquitous reality of human life on earth in 2016 means that where and who you are born to is the most enabling factor in your life. While there is an increasing concentration of wealth in the hands of the 1 per cent and the top 1 per cent of that 1 per cent (the global 0.01 per cent), at the other end of the scale is large-scale debt. Who owns national, family and personal debt? Predominantly the debt of others is an asset for the extremely wealthy. If debt were all quickly paid back, the wealthy would look to lend much of it out again and would be greatly concerned if that became harder to do. And so they make new markets for their lending, encouraging governments to lend university students their tuition fees, for example, rather than pay them from public funds at much lower expense, as used to happen.

Of all the challenges that we have to address in the 21st century, cultural change may be the hardest to accept. Rapid globalisation and the exponential leap forward in digital connectivity, especially within the first two decades of the 21st century, while bringing the world closer together in many respects, have also sharply illuminated fundamental philosophical fault-lines at local, national and international levels. Clashes between cultures have been talked about repeatedly and, of course, that is nothing new. China took a long time to recover from European imperialism, and the indigenous tribes of South America never really recovered from the arrival of the shock troops of the Iberian conquistadores in the 16th century and all the diseases they carried. But now everywhere is being colonised at once by new ideas and a battle between those ideas.

What is most new, from the perspective of so many people today, is the empathetic global outlook now emerging and its increasing influence. We now know that conquistadores, rather than bringing god and salvation, brought disease and purgatory. We now know that the rich take too much, and we understand the environmental consequences of their greed. Activists at all levels have never been more connected, never been so able to share ideas, empathy and solidarity as well as they can today. Surveillance in the digital world is not all one-way: citizens now have the eyes, the means and increasingly the motive to scrutinise. Whistle-blowers such as Chelsea Manning and Edward Snowden demonstrate this.

It is not just the emergence of new globe-spanning ideas but also the images of the digital world that reframe our worldview. People and places come at us with an immediacy that would have been condemned as heretical magic not so many centuries back. You can stroll through suburban Houston on Street View, soar above the Himalaya on Google Earth and tweet from Tahiti. Today we are on the frontline when rioters run amok in Cairo or Kiev and when empathetic Greeks and Italians pluck migrants from the sea. Whether it is using smart phones for banking in Africa or learning new languages from an app in Bangladesh, the mobile phone brings unprecedented access to knowledge, credit and power directly into the hands of many.

Perhaps the only thing that will ultimately threaten the future exponential growth in the dissemination of knowledge and connectivity is a lack of energy to propel it. Whatever connectivity humanity has engaged with over time – whether setting sail by harnessing the wind, toiling across the vast deserts on a camel or even simply sending a text to order a takeaway for dinner – it involves the capturing and the putting to work of energy. Energy will be the key component shaping the map of the future for everybody, everywhere.

When novelists and filmmakers look into the future they tend to be drawn in dystopian directions. Often the state exerts a malign influence and democracy is a far-off dream; prosperity, peace and happiness are consigned to history. In 1949, George Orwell suggested we might inhabit such a world by 1984. The 1982 film *Blade Runner* threw viewers into a future 2019 Los Angeles in a time of climate collapse, when dominating corporations and genetically engineered replicates would build fortresses to protect a few elites from the masses. In the 1999 film *The Matrix*, humans are harvested for energy by sentient machines. More recently David Mitchell, in his 2004 novel *Cloud Atlas*, depicted a future Korea where expendable clones constitute an obedient workforce.

Humans never tire of stories of different possible futures, but not all those stories are dystopian. During the 1960s and 1970s in the television drama *Star Trek*, a confident, successful USA, swept up by its advancements, both in space and through the civil rights movement, imagined a multi-cultural, multi-species world led by a benign Federation. This world operated without money or the need to accumulate possessions. John Lennon sang 'Imagine' at about the same time.

If we are to imagine a positive future what might such a future, built on the principles of sustainable development, actually look like and feel like to live in? In a way the G7 aspires to something that it is unable to define with a clarity that would enable a vision to be communicated to us all. Almost exactly 200 years before this book was published far more utopian ideals could more easily be expressed. On 1 January 1816, a mill-owner, Robert Owen, talked of his wishes for the year 2000:

> What ideas individuals may attach to the term Millennium, I know not; but I know that society may be formed so as to exist without crime, without poverty, with health greatly improved, with little, if any, misery, and with intelligence and happiness increased a hundred-fold; and no obstacle whatsoever intervenes at this moment, except ignorance, to prevent such a state of society from becoming universal.

Perhaps in 2016 we will move closer towards such ideals together? Today we are connected to each other not only by our common relationship with the environment, but also through information sharing and the world of hyper-mobile global capital. Such internationalisation underpins our lives, bringing along a great loss of geographic integrity. What happens on the Chinese stock market affects the value of your pension, irrespective of whether you live in Surbiton or Santa Fe. Many may have noticed this towards the end of 2015 as the Chinese stock market had a substantial downward readjustment. It may even affect whether you ever have a pension if you live in India or Indonesia. It may hasten that day – or hinder it. For humanity, everything is more connected than it has ever been before.

Some suggest a technocratic approach to solving our global dilemmas. It is possible to imagine a technocratic-led world where a re-calibrated capitalism enables greater equality as a result of working in tandem with interventionist governance. This would not necessarily be as pleasant a world to live in as some of its proponents suggest. Technocratic approaches are normally dependent on greater unity between nation-states by means of treaties, regional groupings, international rules, single currencies and even the erosion of national sovereignty, as supra-national organisations such as the EU, IMF, OECD and the UN all gain greater powers.

There has been a long-standing tension between democracy and larger and larger concentrations of power. For technocrats, ideas of 'nudge' theory, benign propaganda and the hegemony of a supposedly especially intelligent elite are possibly more attractive than the unpredictable world of democracy. The battle between the democratic wishes of the Greek people and the European technocrats in Brussels and Frankfurt wishing to decide what is best for those people is a case in point.

One alternative to a world in which a few organisations are increasingly dominant is to become more locally and nationally focused. Although this is not necessarily an either/or choice, in some places movements centred on local identities, narrow nationalism, absolute adherence to a religious code and rural idylls are growing in popularity. Set against the hyper-complex, inter-connected global world, these movements are a retreat to a simpler world, one that is sometimes authoritarian, occasionally liberal and in some cases more libertarian.

For some, rising geographical isolation seems the best strategy in a global competition for increasingly scarce resources. In Britain the UK Independence Party (UKIP) has articulated this position for a British audience; in the USA the Republican Tea Party is its equivalent; and nearly all European nations have had political parties from this part of the ideological spectrum. Often their isolationist thinking comes hand in hand with a rejection of any environmental meta-narrative, such as climate change. In its 2015 general election manifesto, UKIP proposed scrapping the Climate Change Act and reversing the decarbonisation of energy generation by providing support for both coal mining and fracking. The Welsh UKIP MEP Nathan Gill asserted that governments promoted the idea of climate change in order to tax people.

What better frameworks could enable us to address the environmental, social and economic challenges of the future? Can sustainable solutions be achieved? Will sustainability ever even be clearly defined in policy terms?

The Intergovernmental Panel on Climate Change (IPCC) has been charged with modelling various future technological, socio-political and economic possibilities and then predicting the associated environmental impacts, such as by how much sea levels will rise. These scenarios are essentially geo-political projections or, as the IPCC put it: 'Each story line represents different demographic, social, economic, technological and environmental developments, which may be viewed positively by some people and negatively by others.'

The IPCC scenarios span six broad narratives, and although, inevitably, there have been criticisms of the assumptions, methodologies and visioning of these scenarios, they are a relatively effective way of illustrating that socio-political options can lead to substantially different environmental outcomes. The first three stories elaborate on a situation similar to the present with a continuing emphasis on economic growth and increasing economic, cultural and political convergence. Each story differs, however, according to the dominant source of energy utilised. The first is labelled 'Business as Usual' and posits a world where fossil fuels continue to prevail. The second imagines a transition to a mix of fossil and non-fossil fuels – a situation termed 'Greener Business than Usual.' Finally, in the third of these three scenarios, the IPCC models a world of 'Sustainable Globalisation', in which non-fossil fuels are the dominant source of energy.

The IPCC also puts forward three more scenarios significantly different from the current world order. One, termed 'Supra-national Sustainability', corresponds to the ultimate 'think global, act local' position. In other words, this model supposes a world in which sustainable development is the fundamental goal, approached within a framework of overarching global governance. Such a vision is driven by a benign technocratic but human-centred global super-state.

The IPCC considers two further alternatives. These see a greater emphasis placed on national self-reliance, local solutions and the preservation of distinctive local cultures and identities. The key difference between these final two stories is whether their retreats from globalisation are informed by sustainable development or not. In one scenario, 'Deep Green Localism', local sustainability is the focus at the expense of higher levels of economic growth. Some might call this 'steady-state economics', where economies remain at or below their carrying capacity. Conversely, in the 'Competitive Localism' scenario localism becomes more competitive in nature and each place retreats from other cultures. This is a perspective similar to that espoused by UKIP and the Tea Party. Often such thinking is co-joined with social conservatism and the preservation of old orders, structures and beliefs.

Of the six forecasts, the IPCC's 2007 assessment report concluded that the 'Supra-national Sustainability' scenario held the most promise. Under this type of 'super-state' global average annual temperature change during the course of the 21st century has been estimated to be 1.8°C. The 'Deep Green Localism' and 'Sustainable Globalism' action-plans share the silver prize: a rise of 2.4°C is predicted should society follow either of these paths. A rise of 2.8°C is the IPCC's best estimate of temperature change if only 'Greener Business as Usual' is achieved. In contrast, greater rises of 3.4°C are predicted under 'Competitive Localism'. Crucially, the most significant increase (4.0°C) is forecast if 'Business as Usual' prevails.

Dire results emerge when we turn to the IPCC's estimates

for rising sea levels. Under 'Business as Usual', average global sea levels are predicted to rise between 0.26 and 0.59 metres over the current century. This is a greater increase than under any other scenario. If humanity takes the route of 'Competitive Localism' the rise is estimated to be slightly less, between 0.23 and 0.51 metres. In turn, the predicted increase should 'Sustainable Globalisation' or 'Deep Green Localism' be adopted is lower again, at 0.20–0.45 metres or 0.20–0.43 metres, respectively. 'Supra-national Sustainability' takes the gold medal; on this path the world should undergo the smallest rise in global sea levels, of between 0.18 and 0.38 metres.

It is not surprising that these forecasts have been much debated. They are derided by some, and upheld by many others. Recent research tends to suggest that these 2007-based figures may well now be underestimates, yet they remain the best illustration of the relative consequences that these different scenarios could bring about. Indeed, the IPCC's 2007 report was the work of thousands of scientists worldwide, some of them geographers.

Even those parts of the world that will suffer the least from climate change, places like the British Isles and the Korean peninsula, will not be safely cocooned from more extreme weather events, rising sea levels and flooding and the possibility of population pressure from environmental refugees arriving from more afflicted parts of the world.

Bearing in mind that the broadly accepted target is to cap average global temperature rises to 2°C, it is important to imagine what this could mean even in a low-risk area like the British Isles, taking into account not only the geographical impacts but also the adaptations societies would have to make in order to even achieve the 2°C target. Let's therefore end by considering what might transpire under the 'Sustainable Globalisation' scenario in the UK. This is the path that results in, according to the IPCC's best estimates, a 2.4°C average annual global temperature rise over the 21st century.

Imagine the year is now 2100 and that you are living in the UK. First, in terms of the weather, you can expect that 'changeable' will probably be the best description you will hear. Weather records will have been surpassed again and again in recent years, and extreme weather events, such as extensive coastal inundations, with perhaps force 10 storms, might have lashed into the low-lying east coast. The result-ant chaos in the Thames estuary might have led to a prudent but staggeringly expensive upgrade of London's flood defences. Similar ocean forces had churned up the North Sea in 1953, leading to the deaths of over 2,000 people when coastal flooding spread across three English counties. In the 21st century the sea level will be higher, the storms deeper and the winds stronger, but the UK's flood defences could be better.

As the 22nd century dawns, scorching hot summers will be what is most remembered by those looking back: summers of more barbecues and more funerals, as the air quality declines in city centres and heat stress and an ageing population combine to cause a spike in mortality rates. In August 2003 nearly 15,000 excess deaths occurred in Paris as a result of temperatures rising over 40°C. However, people may well have adapted as the 21st century progressed: a mass transfer to electric cars, the phasing-out of diesel engines, huge improvements to public transport and a government-led architectural renaissance creating more compact, energy-efficient cities. Such developments might have alleviated the problem of air pollution as the century progressed, despite temperatures continuing to edge up.

Serious droughts could have been militated against by bringing water supply back under central government control through a single, publicly owned corporation. Most citizens could, by 2100, produce all the electricity they require themselves, and nearly all transport could be powered by electricity. People could still undertake air travel, but perhaps only at the cost of a sizeable portion of their individual 'citizen carbon budget'. A thriving online market, where people could trade their budget allowances, would redistribute income between those who remain determined to fly and those who would never have been able to afford to fly anyway.

Agriculture will have a bumpy ride throughout the course of the 21st century. Some years will be characterised by a lack of rainfall; in other years torrential storms will flush away valuable topsoil into torrid, swollen rivers, bringing devastation that makes those dramatic floods seen in England in 2014 appear like puddles. Food prices will remain high as depleted food supplies elsewhere in the world lead to shortages. People may eat far more seasonally and more healthily in 2100, and consume far less meat and fish.

Britain's population may, by 2050, soar to 80 million, with most of this new population living in apartments in cities; and those cities could produce an increasingly significant proportion of their own food, grown on roofs and green walls and in parks partly turned to allotments. Much of this growth would have been the consequence of migration, possibly partly of environmental refugees fleeing from a flood-ravaged Bangladesh or of a million retirees returning from the Mediterranean and wanting to live in the more pleasant climate that the UK might then have, or a continued influx of the wealthy from all corners of the world. Perhaps the main sources of migrants will be refugees escaping new conflicts and conflagrations across the world, many of which are spawned by competition for resources and global ideological differences. Or perhaps, as normal, they'll be from nearby.

The British economy could be one of the most dynamic, innovative and global in the world if its politicians decided to focus predominantly on raising tax revenue more from wealth than from income, other than high incomes, thereby making most work pay far better than it currently does. There could be EU-negotiated business tax agreements, standardised across Europe, which could drain capital out of the 'shadow economy' through the use of strict tax transparency laws. A proactive European government could have the resources to engage fully with social, economic and environmental sustainability, just as the G7 had said it should do in 2015.

As the IPCC has stated, each of their six scenarios may be viewed positively by some and not by others. What has just been described above is their best-case scenario for a 2°C rise by 2100: a globalised, converging world based on the principles of sustainable development, where less than 5 per cent of all human energy needs are met by carbon-based fuels. But if we do not make informed choices, if disunity reigns at the global level, capping temperature rises at 2°C will be impossible. 'Business as Usual', for example, is estimated to bring increases of 4°C, and the difference between 2°C and 4°C is considerable. It can be measured in millions of additional premature deaths and a rise in social, economic and political pressures that could easily spark more war, create more famine and more disease, both human and pathogens affecting crops and domesticated animals. There is a lot to worry about and many people perhaps need to worry. Geography glues all our worries together.

The challenge for geographers in the 21st century is how to distil and understand the ever-accumulating knowledge and data we have about our planet to ensure the best decision-making for the future. We need to decide what crops to plant to preserve soil fertility and reduce run-off in the long run. We must design better ways to live in more densely settled cities. We should even look to configure a global economic system that works, as Pope Francis states, 'in the service of life, especially human'.

You might be imagining that, given the diversity of academic disciplines and real world stories that fill this account of geography in the 21st century, geography is a subject without a fundamental base. Indeed, geography may appear to cover every subject. Yet, this is exactly what makes geography so important: it is the enabling subject. It enables us to join physics with culture, biology with philosophy, and even zoology with architecture.

Geography is the big picture. It is the subject that studies the accumulation of the deep fertile soils that feed billions. It is about the philosophies we apply to try and understand the world we live in and the people we share it with. It is the inquisitiveness we have in common about our world that matters to all of us.

In a world of dense connectedness – of which we may know an increasing amount – geography enables us to better grasp the complexities and place them within the fundamental framework of our planet – the biosphere that is our ultimate enabler. Geography makes the complex comprehensible, and it provides a context. Geography forces us to look forwards, down the road and into the future. But it looks forwards while also realising that so much that is geographical cannot be understood without looking back.

The processes that have shaped our planet evolved over a long timeframe – the mountains that have been built over multi-millennia, the river valleys carved out gradually and the shorelines that fall and rise with the ages. These processes are almost imperceptible over the course of a human lifetime. The same timeframe applies to the evolutionary forces that have led to the wondrous biodiversity of life. Humanity – although it may be viewed as a highly cognitive species – has uniquely acquired the potential to enact environmental cataclysms upon biodiversity.

Geography focuses on the long term, on both the distant past and far-flung future. It is also a subject of great immediacy. The speed of change currently being enacted upon the Earth caused by humans is of such a scale that an equally rapid response will be required to militate against this human hubris and folly. Never have we been better informed about the multitude of dots – the snippets of information – that make up the big picture. Never before have we had the education and skills to use that information so effectively. Collectively, our ability continues to grow as more and more of the world's marginalised are enabled by education to participate in an informed way about decisions that shape their future.

Geography is, literally, the study of the world. *Geo* is 'earth' and *graphy* is 'writing': geography is 'earth-writing'. The word is as old as the Greek language, and its modern use is as new as the ink on these pages or the pixels that make up the letters you are reading from the screen. Most subjects start with a definition. Geography ends with one: to write about the earth is to write about almost everything we know, everywhere we live and all that we cherish most.

FURTHER EXPLORATION

1 TRADITION

To grasp the importance of a long, global view of economic history in understanding the present geographies of the world, read Andre Gunder Frank, *ReOrient: Global Economy in the Asian Age* (University of California Press, 1998).

A persuasive argument for an early Anthropocene commencing with the first substantial human modification of the environment 7,000 years ago can be found in William Ruddiman, *Plows, Plagues and Petroleum: How Humans Took Control of Climate* (Princeton University Press, 2010).

An excellent undergraduate textbook on the formation of the historical traditional of geography and the many paths that the subject has taken is Anoop Nayak and Alex Jeffrey, *Geographical Thought. An Introduction to Ideas in Human Geography* (Prentice Hall, 2011).

One of the most challenging books written about the fundamental geographical concern with 'place' is Tim Cresswell's *Place: An Introduction* (Wiley-Blackwell, 2nd edition, 2014).

Much lauded, and also often criticised, Jared Diamond's book *Guns, Germs and Steel* (Vintage, 1997) is an attempt to explain the ascendancy of European colonialism from the 16th century onwards.

The original *Tabula Rogeriana* from 1154 is lost to time, but the holder of Oxford University's first Chair of Arabic,

Edward Pococke, mysteriously acquired a 1553 copy in the Middle East in the 17th century. It now resides in the Bodleian Library at the University of Oxford and can be found online at http://en.wikipedia.org/wiki/File:TabulaRogeriana.jpg#/media/File:TabulaRogeriana_upside-down.jpg.

2 GLOBALISATION

Although generally neglected in the British school history curriculum, the Opium Wars are lucidly presented in Julia Lovell's scholarly and brilliantly researched *The Opium War: Drugs, Dreams and the Making of China* (Picador, 2012). Amitav Ghosh's *Ibis* trilogy (John Murray, 2009, 2012 and 2015) brings a master storyteller's eye for detail to this seminal moment in history as he weaves his characters through the worlds of Bengal, Canton, opium and trade in the 1830s and early 1840s. Sometimes the best geographies can be found in fiction.

To understand how the rapid pace of globalisation in the final years of the 20th century started to re-align post-colonial geo-politics and at the same time be entertained by a Pulitzer Prize-winning journalist, read Thomas L. Friedman's *The World is Flat: A Brief History of the Twenty-First Century* (Penguin, 2007).

Trying to understand a rapidly changing China is a complex task, but the subject is made much more access-ible by Martin Jacques's book *When China Rules the World: The End of the Western World and the Birth of a New Global Order* (Penguin, 2nd edition, 2012).

It would be extremely remiss to not mention in any suggested reading about globalisation the work of Joseph Stiglitz, whose book *Globalization and Its Discontents* (Penguin, 2003) has been a milestone in thinking about globalisation.

3 EQUALITY

Richard Wilkinson and Kate Pickett's *The Spirit Level: Why Equality is Better for Everyone* (Penguin, 2010) was first published with the subtitle 'Why More Equal Societies Almost Always Do Better', and many translations have now appeared around the world. It is essential reading, despite already being almost a decade old as far as most of the data is concerned. Its authors specialise in epidemiology, the study of the causes and patterns of disease and health in populations. See also http://www.equalitytrust.org.uk/resources/spirit-level-why-equality-better-everyone.

Among the most cited geographers in the world, although not as cited as Wilkinson and Pickett, is David Harvey. His life's work has concentrated on the injustices and inequalities brought about and reinforced by contemporary capitalism, most recently in *Seventeen Contradictions and the End of Capitalism* (Profile Books, 2014). See also http://davidharvey.org/.

One of the UK's most respected economists is Tony Atkinson (Sir Anthony Barnes Atkinson). In *Inequality: What Can Be Done?* (Harvard University Press, 2015) he not only describes the scale of inequality but sets out a series of measures that could be taken to at least take us back to the lower inequality levels we enjoyed some four decades ago and which much of the rich world enjoys today.

In 2015 a series of charities warned how women in rich countries – particularly in the UK – were especially hurt by spending cuts, which increased inequality between men and women. See, for example, Damien Gayle's article 'Women disproportionately affected by austerity, charities warn', *Guardian*, 28 May 2015; http://www.theguardian.com/society/2015/may/28/women-austerity-charities-cuts-gender-inequality.

The Gapminder Foundation website, developed by Hans Rosling, is not only one of the most effective tools in communicating the historical dynamics of inequality, health and wealth but is also full of entertaining and innovative lectures by Rosling. It can be accessed at http://www.gapminder.org. We particularly recommend the video 'Hans Rosling and the Magic Washing Machine' from March 2011.

4 SUSTAINABILITY

If you want to know how energy use in all that we do and all that is around us is a relative measure, especially in relation to the CO_2 each act generates, Mike Berners-Lee explains all in *How Bad Are Bananas?: The Carbon Footprint of Everything* (Profile Books, 2010).

In 1971 John Rawls published *A Theory of Justice* (Harvard University Press, revised reprint, 1999). This remains one of the most influential books ever written concerning social justice, and social justice is at the very heart of sustainability.

An in-depth investigation into society's awakening to the challenge of climate change has not been more comprehensively covered than by Mike Hulme in *Why We Disagree about Climate Change: Understanding Controversy, Inaction and Opportunity* (Cambridge University Press, 2009).

To turn the clock back to when thinking about sustainability was first launched into a sceptical world, read Rachel Carson's classic *Silent Spring* (Houghton Mifflin, 1962; reprinted as a Penguin Classic 2000).

Green Euro MP and, until recently, Reader in Green Economics at Cardiff University, Molly Scott Cato has produced a highly readable book (an immense achievement for an economics work) evaluating the multiplicity of approaches to economics that concern the environment. See *Environment and Economy* (Routledge, 2011; revised edition 2014).

To see the moment when China finally confronted the continuing pollution catastrophe that its rapid industrialisation has brought to most of its population it is possible to watch all 1 hour and 45 minutes of Chai Jing's documentary 2015 *Under the Dome* with English subtitles at https://www.youtube.com/watch?v=T6X2uwlQGQM.

There is an increasing number of heterodox economists, radical thinkers and sustainability philosophers who are bringing new thinking and ideas to the ecological and economic crisis that beset the 21st century, but revisiting André Gorz and, particularly, his prescient 1991 work *Capitalism, Socialism, Ecology* is a rewarding perspective to consider (Verso, 1994; reprinted 2012).

5 MAPPING THE FUTURE

Some of the most intriguing futurology has been within films and novels. In David Mitchell's novel *The Bone Clocks* (Sceptre, 2014) on 26 October 2043 (well within the lifetime of most readers of this book – even those in middle age or just beyond) we find a world in catastrophic energy decline, where solar panels are the most valuable resource a person can have, the Internet has collapsed, and a corporate China retreats back to its heartland as the one remaining global superpower.

Jeremy Rifkin's *The Empathic Civilization: The Race to Global Consciousness in a World in Crisis* (Polity, 2009) is a more positive perspective on the gifts of humanity and the ability to negotiate the complex challenges of the 21st century.

On allowing in the free-trade jack-boots and all will be well if we have the Transatlantic Trade and Investment Partnership (TTIP), with its 'secretive arbitration panels composed of corporate lawyers, which bypass domestic courts and override the will of parliaments', see: http://www.theguardian.com/commentisfree/2013/dec/02/transatlantic-free-trade-deal-regulation-by-lawyers-eu-us. Alternatively, see: http://www.globaljustice.org.uk/.

Alan Marshall's 'Ecotopia 2121', such as 'Future Car-free Cities across the World', is a good example of work done by more adventurous social scientists; see: http://www.ennr-journal.com/ENRIC/Data/2014/4-Climate%20change%20Vulnerability%208%20papers/PDF/4_15.pdf.

Mariana Mazzucato is Professor of the Economics

of Innovation at the University of Sussex. Her work *The Entrepreneurial State: Debunking Public vs. Private Sector Myths* (Anthem Press, 2013), which was one of the *Financial Times* books of the year for 2013, sets out to challenge the notion that government is the enemy of innovation and dynamism.

Warren Wagar's *A Short History of the Future* (University of Chicago Press, 1989) has been translated into many languages and substantially revised in subsequent editions, showing how hard it is to predict the future even before you get to it so that you can be proved wrong. See: http:// en.wikipedia.org/wiki/A_Short_History_of_the_Future

Finally, for no good reason other than we love it, can we recommend Judith Schalansky's *Atlas of Remote Islands: Fifty Islands I Have Never Visited and Never Will* (Particular Books, 2010)? This book shows where a geographical imagination can take you. Imagine – it's easy if you try.

ENDNOTE

In the medieval European geographical tradition many versions of *Mappa Mundi* were produced – essentially maps of the known world at that time painted on cloth. In the modern version on the following pages, prepared digitally by Benjamin D. Hennig, the land area is proportional to human occupancy and the oceans are nearly eliminated. On such a map it is possible, just for the precise point at which the data were collected, to deduce the central pivot of human population – the centre of humanity at present. As we write, this is a small village, Badi Talai, lying in the arid hills just north of Udaipur, Rajasthan, India. On the world's periphery on this *Mappa Mundi* are some of its currently core economic powers: North America, Europe and Japan.

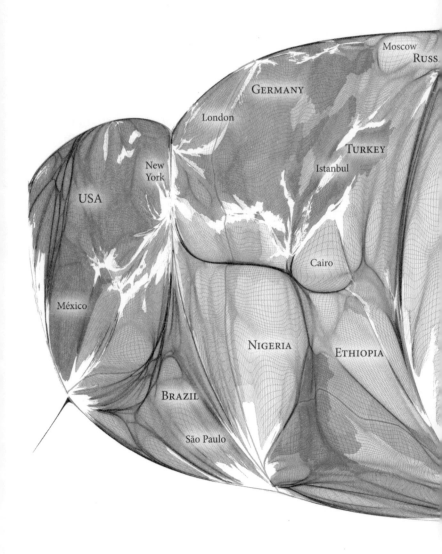

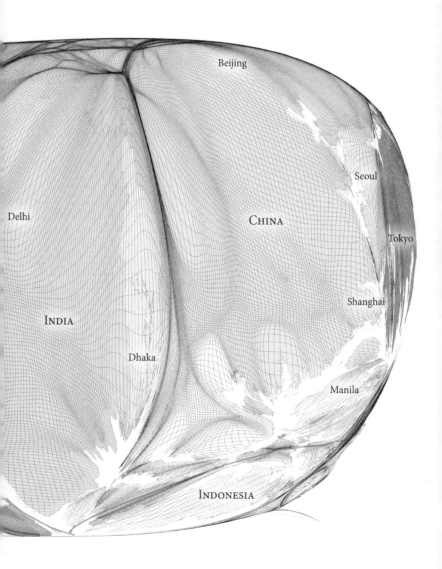

INDEX